BAOFENG RADIO USER GUIDE

A STEP-BY-STEP COMPREHENSIVE MANUAL ON HOW TO USE THE BAOFENG RADIO

ISAAC LEMMINGS

LEGAL NOTICE

The information contained in this book is for information and educational purposes. No part of this book may be duplicated, transmitted, or distributed in any form or by any means, including photocopying, recording, and other mechanical or electronic methods, without the prior written permission of the publisher and the author, except in the case of brief quotations embodied in reviews and other non-commercial uses permitted by copyright law.

Copyright © 2024 **Isaac Lemmings**
All Rights Reserved

TABLE OF CONTENT

LEGAL NOTICE ii

TABLE OF CONTENT iii

INTRODUCTION xi

AN OVERVIEW OF THE BOOK xi

CHAPTER 1 1

   LEGALITIES FIRST 1

      Hazardous Environments 1

      Care and Safety 2

      RF Exposure Information 2

      Warning 2

      GMRS Communication 2

      Licensing Information 2

      Maintenance 3

   Activity 3

CHAPTER 2 4

   GETTING TO KNOW YOUR RADIO 4

      Battery Level Indicator 4

      Status LED 4

      Side key 1- CALL (Broadcast FM and Alarm) 4

      Side Key 2 (Monitor and Flashlight) 4

      VFO/MR - mode key 5

      A/ B select key 5

      Numeric Keypad 5

      Pound # key 5

      Keypad Lock 5

      Star * key 5

      Menu and function keys 5

      Accessory Jack 6

      Power and Volume 6

      Channel selection 6

      Frequency (VFO) mode 6

      Channel (MR) mode 6

      Making a Call 6

   Activity 6

CHAPTER 3 ... 8
NAVIGATING THE RADIO MENU SYSTEM ... 8
Basic Use ... 8
Using Shortcuts ... 8
Making use of the menu shortcuts ... 8
Scanning ... 9
Scanning modes ... 9
Setting scanner mode ... 9
Time Operation ... 9
Carrier Operation ... 9
Search Operation ... 9
Scanning a Frequency Range (VFO mode) ... 9
Scanning Your Chosen Channels (Channel Mode) ... 9
Tone Scanning ... 10
Scanning for CTCSS Tone ... 10
Scanning for a DCS tone ... 10
Dual Watch ... 10
Enabling or disabling Dual Watch mode ... 11
Locking the Dual Watch transmit channel ... 11
DTMF ... 11
Selective calling ... 12
CTCSS ... 12
DCS ... 13
1000HZ, 1450HZ, 1750HZ Tone-burst ... 13
Customization ... 13
Display ... 13
Power-on message ... 14
Activity ... 15
CHAPTER 4 ... 16
REAL RADIO CAPABILITIES ... 16
Baofeng as an FM Radio ... 17
Emergency Frequency ... 17
Activity ... 18
CHAPTER 5 ... 19
UNDERSTANDING VHF & UHF ... 19

- VHF 19
- UHF 19
- Which one is better between VHF and UHF? 20
- Who Uses UHF and VHF? 20
- How Can Signal Strength be Improved on VHF-UHF Two-way Radio? 20
- Activity 21

CHAPTER 6 22
- POWER AND YOUR RADIO 22
  - Charging your Baofeng Radio 22
- Caution 23
  - Charger LED Codes 23
  - Battery Maintenance 23
- Activity 24

CHAPTER 7 25
- ANTENNAS 25
  - How Are Antennas Categorized? 25
  - Main Types of Antennas 26
  - Positioning the Baofeng Antenna 27
- Activity 27

CHAPTER 8 29
- WHAT BAND/FREQUENCY DO I USE? 29
  - FCC RULES 29
- Band Plans 29
  - 2m FM Simplex Frequencies (typical usage, check your local band plan) 30
  - 160 meters (Top Band) 1.800 -2.00 MHz 31
  - 80 meters 3.500 - 4.000MHz 31
  - 40 meters 7.000 - 7.300MHz 32
  - 30 meters 10.100 - 10.150 32
  - 20 meters 14.000 - 14.350 MHz 33
  - 17 meters 18.068 -18.168MHz 34
  - 15 meters 21.000 - 21.450 34
  - 12 meters 24.890 - 24.990 34
  - 10 meter 28.000 - 29.700 34
- Activity 35

CHAPTER 9 36

TALK ABOUT REPEATERS 36

   Types of Repeaters 36

      Telephone repeater 36

      Landline repeaters 36

      Optical Communicator Repeater 36

      Radio Repeater 37

      Broadcast relay station, re-broadcaster, or translator 37

      Microwave relay 37

      Passive repeater 37

      Cellular repeater 37

      Amateur radio repeater 37

   Using a Repeater 38

      Finding a Repeater 39

      Operating a Repeater 39

      First Transmission 39

      An autopatch 40

   Activity 41

CHAPTER 10 42

   RADIO NETS AND FINDING THEM 42

   The basic Structure of the Net 42

   Modes of Net Operation 43

      Directed Net 43

      Free net 43

      Amateur Radio Net 43

      Amateur Net Operation 44

      Formal Operation 44

      Informal Operation 44

   Types of Amateur Nets 44

      Traffic 44

      DX 45

      Club or Topic 45

   Finding and Participating in Ham Radio Nets 45

   Checking in with Ham Radio Nets 46

   Activity 46

CHAPTER 11 48

PROGRAMMING WITH CHIRP 48
   Preparing to Program Your Baofeng Radio 49
      Alternative options 50
   Weather stations 50
   Local repeaters 51
   Emergency and first responders' frequencies 51
   FRMS AND GRMS 52
   Activity 52
CHAPTER 12 53
   HAND PROGRAMMING THE BAOFENG 53
   Programming Simplex Channels into Baofeng 53
   Programming Repeaters into a Baofeng 54
   Deleting a Channel 56
   Theory of Operation 56
   Activity 56
CHAPTER 13 57
   LINGO & TALKING ON THE RADIO 57
   Q codes used by Ham Radio Operators 58
      QRL 58
      QSL 58
      QRP 58
      QRU 58
      QRS 59
      QRV 59
      QRT 59
      QSO 59
      QRZ 59
      QTH 59
   Communication Terms 59
      73 59
      88 60
      DX 60
      99 60
      XYL 60
      YL 60

  COPY 60

 Phonetic Alphabet.. 60

 10 Codes... 61

 Activity 62

CHAPTER 14 63

 FINDING A CLUB 63

 Reasons to Join a Radio Club 64

 Activity 65

CHAPTER 15 66

 BASICS ABOUT COAX & YOUR BAOFENG 66

 Types of Coaxial Cable 67

 Coaxial Cable and Baofeng... 68

  RG-6 Cables 68

  RG-59 Cables 68

  RG-11 Cables 68

 Activity 69

CHAPTER 16 70

 ARE YOU GOING TO USE THAT RADIO? 70

  Cost of Ownership... 70

  Enhanced worker safety 71

  Battery Life and Management 71

  Durability and Reliability... 71

  Audio Quality 71

  Value-adding Features and Functionality 71

 Activity 72

CHAPTER 17 73

 NON-HAM COMMUNICATIONS 73

 Multi-Use Radio Service (MURS).. 73

 General Mobile Radio Service.. 73

 Family Radio Service 74

 Citizens' Band Radio... 74

 Activity 76

CHAPTER 18 77

 GETTING LICENSED & LEGAL... 77

 Getting Your Ham Radio License.. 77

  Choose your level of license 77
    Technician License 78
    General License 78
    Amateur Extra License 78
  GMRS License 78
  Activity 79
CHAPTER 19 80
  SOME RECOMMENDED ACCESSORIES 80
  Activity 81
CHAPTER 20 82
  A LOOK AT WINLINK 82
  Winlink Express 82
  Message Entry 84
    Message Sending/ Receiving Sessions 84
  Ham Operators and Winlink 84
  Winlink Account 85
    Make use of a different installation process apart from Winlink Express 85
  Activity 85
  Conclusion 86
  Appendix A: Calling Frequencies 87
    160 METERS 87
    80/75 METERS 87
    40 METERS 87
    30 METERS 88
    20 METERS 88
    17 METERS 88
    15 METERS 88
    10 METERS 89
  Appendix B: Bands for LEGAL Use 89
  Appendix C: GMRS/FRS Frequencies & Data 90
  Appendix D: MURS Frequencies 91
  Appendix E: VHF Marine Frequencies 92
  Appendix F: Frequency Coordinators 95
  Appendix G: SWR Meters 98
  Appendix H: CTCSS Squelch Tones (Hz) 98

Appendix I: DCS Codes 99

Appendix J: Phonetic Alphabet 99

Appendix K: Radio Horizon Antenna Heights 99

Appendix L: Weather Radio Frequencies 99

INDEX 101

INTRODUCTION

A Chinese firm of the same name produces the portable radio known as Baofeng. This model was the first dual-band VHF/UHF radio made in China to be widely distributed internationally. Due to its low cost and simplicity of use, amateur radio amateurs and professionals use it extensively.

It can transmit on the 2m band, which is between 136 and 174 MHz (VHF) (UHF), and the 70cm band, which is between 400 and 520 MHz (UHF). Among the features that come standard on this two-way radio are dual watch and dual reception, an LED flashlight, voice instructions in Mandarin or English, and adjustable LED illumination for the LCD. Channels and frequencies must be manually coded, which takes time. To enhance your radio experience, computer-based programming is made simple with the use of CHIRP software and a specialized USB-to-radio adaptor.

AN OVERVIEW OF THE BOOK

This book has been carefully put together to provide you with all the information you need to get started with the Baofeng radio and make the most of it. The twenty chapters in this unique book are broken down into the following snippets from each of those chapters.

Chapter 1: Legalities First
This chapter provides a detailed explanation of all the legal requirements needed to operate the Baofeng radio. Here, you will learn whether using the Baofeng radio is permitted in your nation as well as the names of the regulatory bodies that oversee its usage.

Chapter 2: Getting to Know Your Radio

You will learn more about your radio in this chapter. Along with learning how to operate the numeric keypad, side keys, battery level indicator, and radio to make phone calls, you will also learn about the status LED.

Chapter 3: Navigating the Radio Menu System

Everything related to the radio menu will be covered in this chapter. Along with learning how to use shortcuts and the many scanning modes, you will also learn how to personalize the radio.

Chapter 4: Real Radio Capabilities
This chapter will enlighten you on the various capabilities of the radio as well as how you can make use of the radio as an FM radio.

Chapter 5: Understanding VHF & UHF
This chapter will provide you with additional information about VHF and UHF, help you choose which one is ideal for your needs, and show you how to boost both frequencies' signal strengths.

Chapter 6: Power and Your Radio

Without power, your radio will not be active. In this chapter, you will learn to charge the Baofeng radio, understand the charger LED codes, and also be enlightened about how to maintain your battery.

Chapter 7: Antennas

Antennas contribute to clearer transmission. The primary categories of antennae and the correct placement of the Baofeng antenna are covered in this chapter.

Chapter 8: What Frequency Do I Use?

In this chapter, you will learn about the various frequencies you can use with the Baofeng radio.

Chapter 9: Talk about Repeaters

In this chapter, you will be enlightened about what repeaters are, how they can be used, and the various types of repeaters. You will also learn how to locate a repeater.

Chapter 10: Radio Nets and Finding Them

In this chapter, you will get to know what radio nets are, the modes of net operation, what amateur radio nets are, and also the types of amateur nets.

Chapter 11: Programming with CHIRP

In this chapter, you will learn about CHIRP and how to use it in the programming of your radio.

Chapter 12: Hand Programming the Baofeng

There are other ways to program the Baofeng than using CHIRP; in this chapter, you will discover how to program the Baofeng manually.

Chapter 13: Lingo & Talking on the Radio

Speaking on your smartphone and speaking on the radio are two very different things. This chapter will teach you the basics of using a radio to communicate as well as the different codes that you can use.

Chapter 14: Finding a Club

No one knows it all, and being in a group can further propel you to attain greater heights. In this chapter, you will understand the reason why you need to join a club of ham operators.

Chapter 15: Basics about Coax & Your Baofeng

When it comes to using radios, coax or coaxial cables are crucial. You will discover more about this cable in this chapter, including its different types and which ones work with your Baofeng radio.

Chapter 16: Are you going to use that Radio?

This chapter is quite interesting as it simply explains the benefits of using the Baofeng radio.

Chapter 17: Non-Ham Communications

Not all radios are Ham radios. This chapter brings a list of radios that can also be used for communication but are not Ham radios.

Chapter 18: Getting Licensed & Legal

Most people undoubtedly wish to avoid running afoul of the law. Using your Baofeng radio to stay legal is made easier with the guidance of this chapter.

Chapter 19: Some Recommended Accessories

In this chapter, you will learn about the various accessories you can make use of with your Baofeng radio.

Chapter 20: A Look at Winlink

In this chapter, you will learn about Winlink and its importance as regards radio communication.

CHAPTER 1
LEGALITIES FIRST

Given that Baofengs are widely available on any frequency between 136 and 174 MHz and 400 and 520 MHz, it is unlawful to import, promote, or sell them. Nonetheless, having or using a Baofeng does not violate the law if you are a licensed amateur radio operator and exclusively use the equipment on amateur radio frequencies. The only way to utilize one of these "illegal" pieces of equipment outside of amateur bands is to violate amateur licensing laws.

This device complies with the FCC rules (Part 15). The operation is subjected to the following conditions;

- This device might not lead to harmful interference.

- This device must allow any form of interference received which includes interference that might lead to an unwanted operation.

Any alteration or modification not expressly authorized by the organization in charge of compliance may revoke the user's authorization to use this device. Remember that this particular radio is programmed to transmit a regulated signal on a designated frequency. Furthermore, be aware that it is illegal to alter the radio's settings in an attempt to surpass the pre-established limitations. Any adjustments to the radio must be made by technicians who hold the required credentials.

To be safe and sure:

- Ensure you never open the case of your radio.

- Do not change anything in the radio components. Although, you are at liberty to change the battery of the radio if need be.

When your television or radio is working correctly, your radio may produce interference. All you need to do to find out whether this is occurring is switch off your radio and see if the signal that the TV or radio is receiving changes. Nevertheless, by relocating your radio away from the TV or radio, you can get rid of this interference. The Federal Communications Commission recommends that you stop using the radio if you are still unable to remove this interference.

Hazardous Environments

Make sure the radio is not used in dangerous situations. If this occurs, there could be an explosion or fire. Additionally, avoid working near electrical blasting caps that are not insulated.

Radio interference can occasionally disrupt blasting operations, which increases the risk of an explosion. In locations where there is a "Turn off two-way radio" sign or if there is blasting, turn off your radio to avoid unintentional transmission. Construction workers frequently utilize a remote-control gadget to detonate explosives.

Care and Safety
A gentle cloth moistened with a small bit of water can be used to clean the radio. Make sure not to use any kind of solvent or cleanser as they could damage the radio's body and generate leaks that could cause irreversible damage. Use a dry, lint-free cloth to clean the battery's contacts.

RF Exposure Information
The United States Federal Communications Commission (FCC) amended and adopted the safety standard for human exposure to radiofrequency electromagnetic energy radiated by FCC-regulated transmitters in August 1996 with its action in Report Order FCC 96-326. These recommendations are thought to be in keeping with the safety standards that have previously been established by international and U.S. standards organizations. The radio's design complies with various international requirements as well as FCC norms.

Never allow a child to operate a radio without an adult's close supervision and understanding of the following recommendations;

Warning
The user of the radio has the sole responsibility of ensuring that the radio transmitter is used safely. Please ensure strict adherence to the instructions below;

- Use the only antenna that is provided or authorized. Unauthorized antennas, modifications of any kind, and attachments may deteriorate the quality of the call, harm the radio, or result in an FCC violation.

- Make sure no broken antenna is used with the radio.

- An antenna that is damaged could burn the skin if it comes into touch with it. If you ever feel that your antenna isn't working properly, please get in touch with your local dealer to get it replaced.

GMRS Communication
This land-mobile function, known as GMRS (General Mobile Radio Service), is available for two-way, short-distance communications within the United States. To communicate smoothly on the frequencies on the channel that is available on this radio, you must possess a current FCC license. The GMRS/FRS frequencies that this radio uses are designated for use while working, hiking, keeping in touch with family and friends during a busy public event, checking in with traveling companions in a separate vehicle, conversing with neighbors, and scheduling meetings with other people while shopping in the mall. The FCC will issue a call sign to licensed users, which they must use to identify their station when using this radio. Additionally, GMRS users should make sure they cooperate by only participating in authorized transmissions, staying out of other GMRS users' channels, and using caution when extending their transmission times.

Licensing Information
The Federal Communications Commission (FCC) license is required for this two-way operation on GMRS frequencies. Before using this radio for transmission on the GMRS band, a user must hold a license. In breach of FCC regulations, the unauthorized use of GMRS channels may

result in severe fines. Additional guidelines for using this radio can be found in 47 C.F.R. Part 95. To obtain application forms and licensing information, simply contact the FCC Hotline. Further information can be found at www.fcc.gov, the FCC's website.

Note: You must possess an FCC license even if you use this radio at considerably lesser power (0.5 watts) on FRS (Family Radio Service) channels. All GMRS regulations are applicable because the radio runs in the 0.5 to 5-watt GMRS power range, and you must have a GMRS license to use the FRS (Family Radio Service) feature. Typical FRS-only radios have an integrated (non-detachable) antenna and can run at a maximum power of 0.5 watts (500 milliwatts).

Maintenance

Your two-way radio is an electronic product of exact design and ought to be treated with utmost care.

The instructions below will aid you in the fulfillment of any warranty obligations and also in enjoying the radio for many years.

- For whatever reason, do not attempt to open the radio. For the same reason that the radio's electronics and precise mechanics require training and specialized tools, the radio should never be realigned because it has already been calibrated for the best performance. Once the radio has been opened without authorization, the warranty is invalidated.

- Make sure the radio is not kept in direct sunlight or a heated environment.

- Elevated temperatures have the potential to shorten the device's lifespan by melting or distorting some plastic materials.

- Ensure the radio is not kept in dusty places.

- Ensure the radio is kept dry. Rainwater or dampness will corrode electronic circuits.

Activity

1. Make a brief description of the legal steps that should be taken to make use of the Baofeng radio legally.

CHAPTER 2
GETTING TO KNOW YOUR RADIO

Below are the various tips you must take note of at first as having to do with your radio;

- Before using the radio, make sure the antenna is correctly installed and the battery is fully charged.

- Try not to tighten your antenna over to prevent harm to the connection base and outside materials.

- Recall to hold the antenna by its base and screw it in place when installing it.

- If you decide to utilize an external antenna, make sure its SWR is at least 1:5:1 to prevent damaging the transceiver.

- Never try to transmit anything without the usage of an antenna.

Battery Level Indicator
When the battery level indicators display an empty battery. It simply indicates that the battery is low on energy and has to be charged right away. At this point, the radio will start beeping periodically and flashing the display's lights. It will also start to declare "Low Voltage" audibly if voice prompts are enabled, indicating that you need to charge your battery.

Status LED
The status led has a very simple and quite traditional design. Whenever you get a signal it displays the green light, when you make a transmission; the red light is shown and when the radio is on standby, the LED light will be off completely.

Side key 1- CALL (Broadcast FM and Alarm)
Press the Call symbol on the radio for a brief while to start the FM broadcast reception. Pressing again for a brief moment will switch off the broadcast FM receiver. The squelch will be activated to that specific frequency as if you were doing a scan if you happen to receive a signal on the active channel or frequency while you are still listening to broadcast FM. It will stay there until the signal leaves, at which point it will return to broadcast FM.

Tap and hold the call symbol to turn on the radio's alarm feature. You only need to give the call button another quick press to stop the alarm from going off.

Side Key 2 (Monitor and Flashlight)
To turn on the LED flashlight, briefly tap the Moni symbol. A further brief push will cause the LED to flash. Pressing the button one more time will completely turn off the flashlight. To keep an eye on the signal, tap and hold Moni. This will allow the squelch to take over and allow you to hear the unfiltered signal.

VFO/MR - mode key
You can switch between the frequency (VFO) and memory (MR) modes by tapping VFO/MR. Memory mode is also known as channel mode. You must first be in the Frequency (VFO) mode if you wish to save frequencies to channel memory.

A/ B select key
The A/ B key changes between A (upper) and B (lower) displays. The frequency or channel on the chosen display will then be the active listening and transmit frequency or channel as the case may be.

Numeric Keypad
There is a typical complete numeric keyboard included with the Baofeng handheld receiver. Their secondary purpose is embedded in these keys. On the other hand, there is a real secondary purpose for the 0 and # keys, which is to locate and scan the device.

Pound # key
The # key serves as a transmit power shift key in channel mode as well. Press and hold the # key to switch between High and Low transmit power while in channel mode. Make sure that this just affects the active session and has no effect on the transmit power saved to memory for that particular channel. The transmit power will be reset to what is stored in channel memory if you switch to a different channel or operating mode (which includes broadcast FM).

Keypad Lock
The Baofeng has a keypad lock feature that helps to lock all keys except for the three keys at the side of the radio.

If you would like to either enable or disable the keypad lock, tap and hold the key for about two seconds. You can also enable the keyboard so that the radio will lock the keypad automatically after about ten seconds from the menu.

Star * key
Pressing the key for a little moment enables the opposite function. Pressing the button briefly will start the scanning process when you need to listen to broadcast FM. Regardless of the scanner resume mechanism, scanning in broadcast FM will stop as soon as an active station is found.

To activate the scanner, hold down the Scan key for approximately two seconds.

Menu and function keys
Typically, the menu key is used to confirm menu items and enter the menu. In addition to selecting channels and stepping up or down in frequency depending on the mode of operation, the upper and downward keys are utilized to navigate through the menu. You can cancel menu options and exit menus by using the exit key.

Accessory Jack

The accessory jack on the Baofeng can be described as a Kenwood-compatible two (2) - pin design.

Power and Volume

You can turn the radio on by turning the volume/power knob clockwise until you hear a clicking sound. When your radio turns on properly, it should sound like a loud double beep after a moment or so. After that, it should display a message or flash the LCD for a minute or so, depending on your settings. It will then display a frequency or channel after that. The voice will proclaim "frequency mode" or "channel mode" if the voice prompt is enabled. Simply flip the volume/power knob counterclockwise until you hear a click to turn the device off. Then the radio will shut off. You can adjust the radio's volume by rotating the volume/power knob clockwise.

Turn the volume/power knob counterclockwise to lower the volume. Be careful not to twist the knob too far, since you can unintentionally turn off the radio.

Channel selection

There are two distinct operating modes: Channel or Memory (MR) mode and Frequency (VFO) mode. The Channel (MR) mode will be far more useful than the frequency (VFO) option for everyday use. However, channel programming into memory can also be done using the Frequency (VFO) mode. The mode you end up using will ultimately depend on your use case.

Frequency (VFO) mode

When in frequency (VFO) mode, you can use the upward and downward keys to traverse up or down the band. Depending on the frequency step the transceiver has been set to, there will be a frequency increment or decrease with each press.

Channel (MR) mode

The use of the Channel (MR) mode is based on basically having programmed certain channels to make use of. Once the channels have been programmed and ready, you can then make use of the upward and downward keys to move between different channels.

Making a Call

The PTT button on the side of the radio's body is where you tap and hold to start a call with the Baofeng radio. As long as the transmission is in progress, make sure you are speaking 3–5 cm away from the microphone. Your transceiver will switch back to receive mode when you release the PTT button. Additionally, you can directly infer frequencies with kilohertz accuracy on your numeric keypad. However, the radio will floor to the next frequency that matches your frequency step. This simply implies that you should always round up your input when you input frequencies that are higher than 1 kHz resolution, such as 145.6875 MHZ.

Activity

1. Study the battery level indicator to take note of what the various lights indicate.

2. Make use of the side keys of your radio.
3. Enter keys with the use of the numeric keypad.
4. Make a phone call with the use of your radio.

CHAPTER 3
NAVIGATING THE RADIO MENU SYSTEM

Basic Use
- Tap the **Menu key to enter the menu.**
- Make use of the downward and the **upward keys to move between menu items.**
- Once you've found the menu item you want to select, tap **it once more.**
- To select the desired parameter, use t**he upper and downward keys.**
- Once you have selected the parameter you want to set for a certain menu item, press the menu button to confirm your choice. This will save your settings and return you to the menu.
- You can exit your menu and reset that menu item by tapping the menu button if you want to undo your modifications.
- To exit the menu whenever you'd like. Press **the button.**

Using Shortcuts
As you may have noticed, each item on the menu has a numerical value associated with it. You can use these numbers to get straight to any item on the menu.

Additionally, the menu is set up so that the ten most often used functions are at the top and are printed on the keypad so you don't have to memorize them all. There's also a number associated with the parameters.

Making use of the menu shortcuts
- Tap the **menu key to enter the menu.**
- Make **use of the numerical keypad to enter the number of the menu item.**
- Tap the **menu key to enter the menu item.**
- **If you would like to insert the preferred parameter, there are about two options available;**
    - Make use of the arrow keys or make use of the numerical keypad to insert the numerical short-cut code.
    - To make a confirmation of your selection, tap **the menu button** and it will save your setting and bring you back to the main menu.
- To cancel your changes, tap **the exit button** and it will reset that menu item and bring you out of the menu completely.

- To leave the menu at any time, tap **the exit key.**

Scanning

A VHF and UHF band scanner is integrated inside the Baofeng. It will scan in stages according to the frequency step you have selected while it is in frequency (VFO) mode. It will scan your channels when in Channel (MR) mode. Typically, Dual Watch is disabled while scanning

Press and hold the scan key for approximately two seconds to turn on the scanner. To exit the scanning mode, tap any key.

Scanning modes
The scanner can be configured to any of the three ways of operation; Time, carrier, or search, each of which has been explained in the section below;

Setting scanner mode
- Tap the **menu key to get into the menu.**
- Enter **Step on your numeric keypad to switch to scanner mode.**
- Tap the **Menu button to make a selection.**
- Make use of the downward and upward arrow keys to choose your preferred scanning mode.
- Tap the **menu key to confirm and save.**
- Tap the **exit key to leave the menu.**

Time Operation
In Time Operation (TO) mode, the scanner will stop scanning when a signal is detected and start again after a factory-set timeout.

Carrier Operation
When a signal is detected in carrier operation (CO) mode, the scanner stops and waits for a certain amount of time before continuing to scan.

Search Operation
In Search Operation (SE) mode, the scanner will stop when a signal is detected. If you would like to resume scanning, press and hold the **Scan key** once more.

Scanning a Frequency Range (VFO mode)
A user-specified frequency range can be scanned by Baofeng radios. For approximately two seconds, tap and hold the scan button. After that, the radio will begin scanning the frequency.

Scanning Your Chosen Channels (Channel Mode)
Baofeng can scan the channels in your programmed memory; you can easily add or remove channels from the scanning bank at any time. To start scanning, tap and hold the scan button

for approximately two seconds. In the scan cycle, channels with a star icon next to the channel number will be added.

Tone Scanning

While selecting either Channel Mode (MR) or Frequency Mode (VFO), it is possible to scan for a CTCSS tone or DCS code. The detected tone or code is only recorded to the menu 11/10 when the VFO mode is selected.

It is possible to access the CTCSS tone and DCS code scanning mode both with and without a signal. The actual scanning procedure will only start when a signal is received. It's important to remember that not every repeater that accepts a DCS code or CTCSS tone for access will send one back. In that scenario, it would be necessary to scan the transmitter of any station that has access to the repeater. This merely indicates that stations on the repeater's input frequency would be listened to to do this.

Scanning for CTCSS Tone

- Press the menu key to enter the menu.
- Enter the step button on your numeric keypad to come to menu 11
- Tap the menu key to select.
- Tap the scan button to commence scanning.

The radio will be in CTCSS scanning mode when a flashing CT appears on the left status display. In this mode, the bottom display will cycle through the CTCSS tones while they are being tested whenever the radio detects an RF signal on the selected MR channel or VFO frequency. After determining the frequency of the received CTCSS tone, the "CT" indicator will stop blinking.

Scanning for a DCS tone

- Tap the menu key to enter the menu.
- Enter 10 on your numeric keypad to get to Menu 10.
- Tap the menu key to select.
- Tap the scan button to commence DCS scanning.

The left status display will have a flashing "DCS" to indicate that the radio is in DCS scanning mode. In this mode, the lower display will cycle through the DCS codes while they are being tested whenever the radio detects an RF signal on the selected MR channel or VFO frequency. The "DCS" indication will stop blinking after the bits of the DCS code have been received and calculated. To save the scanned tone in memory (VFO Mode Only), press the menu key. To depart, press the exit key.

Dual Watch

The capacity to communicate with two channels at once can prove to be a highly useful tool in some circumstances. With its Dual watch feature, the Baofeng may lock the transmit frequency to one of the two channels it monitors and alternate between two frequencies at predetermined intervals to enhance its functionality.

Enabling or disabling Dual Watch mode

- Press **the menu key** to enter the menu.
- Enter **7 on the numeric keypad** to get to Dual Watch.
- Press **the menu button** to choose it.
- Make use of the downward and upward keys to either enable or disable.
- Tap **the exit key** to leave the menu.

The Baofeng radio is designed in a way that when any of the A or B Frequencies (VFO/MR) becomes active, it will automatically start transmitting on that channel. This behavior can be exceedingly annoying, particularly if you are listening in on a frequency that you shouldn't be using for transmission. The transmitter can be locked to either the A or B channel via a menu option.

Locking the Dual Watch transmit channel

- Press **the menu key** to enter the menu.
- Enter **34** on the numeric keypad.
- Tap **the menu key** to select.
- Make use of the upward and downward keys to choose the A (upper) or B (lower) display.
- Press **the menu key** to confirm it.
- Press **the exit key** to leave the menu.

DTMF

Dual sinusoidal signals are used in this in-band signaling technique for each given coding. Created primarily for telephone networks, it has proven to be an extremely useful tool in most other fields as well.

DTMF is frequently used for automation systems and remote control in two-way radio systems. Amateur radio repeaters are a well-known example, where certain repeaters are turned on by transmitting a DTMF sequence, which is typically just a straightforward one-digit sequence.

The A, B, C, and D codes are all included in the complete implementation of Baofeng radio. The DTMF codes A, B, C, and D are found in the menu key, upward and downward keys, and exit keys, respectively. The number keys and scan keys also correlate to the corresponding DTMF codes. By holding down the PTT key, hit the keys that correspond to the message you want to transmit to send DTMF codes. It should be noted that you do not need to unlock your radio to send DTMF tones if your radio's keypad is enabled.

Selective calling

When working with bigger groups of individuals over the same channel, there are occasions when the conversation can get a little chaotic. Many techniques for filtering out unwanted transmissions on your frequency have been created if you would like to lessen this issue. Group calling and individual calling are the two main types of selective calling in two-way radio systems.

Group calling can be characterized as a one-to-many communication method, as the name suggests. Each radio in your working group is configured in the same manner, and all of the radios in the group will communicate with each other.

Individual calling, sometimes also known as paging can be described as a one-to-one mode of communication. All the radio is programmed with a special ID code and only by sending out a matching code will you be able to get that radio to open up to your transmissions.

Most Baofengs have about three different ways of group calling;

- CTCSS
- DCS
- Tone-burst

CTCSS
CTCSS is configured with menus 11 R-CTCS and 13 T-CTCS.

Procedure

- Press **the menu key** to enter the menu.
- Enter **11 on the numeric keypad** to get to the receiver CTCSS.
- Tap **the menu key** to select.
- Insert **the preferred CTCSS sub-tone** frequency in hertz on the keypad.
- Tap **the menu key** to confirm and save.
- Enter **13** on the numeric keypad to go to transmitter CTCSS.
- Tap **the menu button** to select.
- Insert the preferred CTCSS sub-tone frequency in hertz on the numeric keypad. Ensure it is the same frequency as the one you inserted for receiver CTCSS.
- Tap the **menu button to confirm and save.**
- Tap **exit to leave the menu system.**

If you would like to switch off the CTCSS, follow the exact procedure but ensure it is configured to off with the 0 SQL key rather than having to choose a CTCSS sub-tone frequency.

DCS
This is configured with menu 10 R-DCS and 12 T-DCS

- Press **the menu key** to enter the menu.
- Insert **10 on the numeric keypad** to move to receiver DCS.
- Tap **the menu key** to select.
- Insert **the preferred DCS code** on the numeric keypad.
- Tap **the menu button** to confirm and save.
- Insert **12 on the numeric keypad** to go to the transmitter DCS.
- Tap **the menu button** to select.
- Insert **the preferred DCS code** on the numeric keypad. Ensure it is the same code as the one you entered for receiver DCS.
- Tap **the menu button** to confirm and save.

1000HZ, 1450HZ, 1750HZ Tone-burst

All you have to do to send out a tone burst is to hold down the PTT and simultaneously push a key. Using this functionality essentially requires no configuration.

Below are the configurations that will transmit accordingly;

PTT + the call button = Transmits 1000 Hz Tone Burst

PTT + VFO/MR = Transmits 1450Hz Tone Burst

PTT + A/B = Transmits 1750Hz Tone Burst

Customization

With Baofeng radios, you can personalize the power-on message and the illumination color in each of the transceiver's three states (transmit, receive, and standby).

Display

The LCD on the Baofeng radio is backlit by multi-color LEDs, the color of which can also be pre-set from the menu system into diverse colors.

If you would like to modify the colors, take the steps below;

Changing the backlight color

- To access the menu, tap the menu key.
- Enter a number on your numeric keypad from the following.
- To change the standby color, **insert 29.**

- To change the received color, insert **30.**
- To change the transmit color, insert **31.**
- Press **the menu ke**y to select.
- Make use of the upward and downward keys to choose your preferred color.
- Tap **the menu ke**y to confirm and save.
- Press **the exit button** to leave the menu.

To modify the duration of the backlight for your LCD follow the steps below;

Configuring backlight time-out

- Tap **the menu key** to enter the menu.
- Tap **6 on your numeric keypad** to come to backlight time out.
- Press **the menu key** to select.
- Make use of the upward and downward keys to choose the preferred color.
- Press **the menu button** to confirm and save.
- Tap **the exit button** to leave the menu.

Syncing the Display

- To access the menu, tap **the menu key.**
- On your numeric keypad, enter **42 to access the Sync Menu.**
- To choose, use the menu key.
- To select "**ON,**" use the upper and downward keys.
- Press the menu button to save and confirm.
- To quit the menu, tap **the exit button.**

Power-on message

Only the computer link can be used to configure the power-on message. The following instructions are predicated on the supposition that the Baofeng program has been installed and is operational and that a link has been created using it from a Windows machine.

Configuring the power-on message

- Choose "other" from the menu bar to launch the "other" conversation.
- The two text fields in the "Power on Message" box correspond to the two lines on your LCD. Enter the desired text in the designated fields.
- To write your changes to the radio, select Write.

Activity

1. Make use of the various scanning modes in your Radio.
2. Scan a frequency range.
3. Scan a channel.
4. Customize your radio; modify the display as well as the power-on message.

CHAPTER 4
REAL RADIO CAPABILITIES

For any activity to be completed properly, communication is essential. Being able to speak well could mean the difference between life and death in a crisis. Because of this, it is a crucial component of safety.

You are usually in places without mobile coverage when you are out prospecting. Thus, you require a different means of communication. Because they are portable, handheld radios are the ideal choice for interpersonal communication. Using a radio is better because of its field-ruggedness and longer battery life, even if you are within cell phone coverage.

Made in China, the Baofeng radio is a dual-band amateur radio. It is meant for those who are new to it. These radios are available on Amazon.com for less than thirty bucks per! Larger and occasionally even smaller commercial entities would use far more expensive radios—like the Kenwood TK-3402—to connect with their employees and clients. Compared to other devices, these radios are far less feature-rich and retail for more than $300. To have your Kenwood radio programmed, you will typically need to take it to a dealer. Almost everyone brings them to a dealer to get them programmed, but if you have the right cable and software, you can do it yourself.

There is a hidden gem in Baofeng. Many of them had been using significantly more expensive radios for years, so their expectations were low for a Chinese set that cost less than $30. They were surprised to learn about the Baofeng radio's capabilities. The most noteworthy and possibly most helpful aspect of Baofeng is its on-the-spot radio frequency programming capability. You will notice the frequency marked at the start of the road when you go to a road that is either actively being logged or maintained by the British Columbia Forest Service. Furthermore, as you move up the road, you should proclaim the kilometers. You want to do this because enormous logging trucks and other equipment are working up there, and you want to avoid getting in their way. If you can converse with them, you will be able to avoid situations where you could be run over by logging trucks or become trapped on narrow roads.

You'll be wondering why field programming isn't available on other radios. This is because, in terms of the law, transmitting across several channels requires a license. Using the Baofeng radio could get you into a lot of trouble because you can adjust it to receive whatever channel you want. Listening in on the conversations between ambulances and police is not hard. Sometimes people listen in on other people's conversations, even those with the police, as a kind of amusement. It is also possible for you to send information unlawfully. Having said that, in the event of an emergency, it would be prudent to contact for help independently.

It's also reported that this radio has an incredible range. Messages may be sent at approximately 4 watts, while commercial-grade Kenwoods have a 5-watt output. The majority of users have tested the Baofeng radios' range at over 10km, and with a clear line of sight, they may be able to reach considerably farther. Additionally, there's a dual watch capability that lets you simultaneously monitor two separate channels. The last channel with activity will have a transmission when you press the PTT button. The radio features a scan feature as well, however it is very slow.

The factory battery has a life expectancy of up to 20 hours. That is a respectable amount of time; one would not anticipate any other radio to function for a longer period. You also have the option to buy replacement batteries, which can be done for a price that is not prohibitively expensive. These may be purchased at a price of approximately $6.00 each. The VOX capability is something that is typically only accessible on radios that are significantly more expensive than the Baofeng model. VOX enables you to control the PTT with just your voice, allowing for virtually hands-free operation of the device.

Programming the Baofeng to work with repeaters—like the ones the BC Forestry Service uses, for instance—is quite easy. This feature helps make the radio more versatile as a tool for communicating during emergencies. Programming on a mobile device can occasionally be a little bit difficult, but it is not impossible. A personal computer (PC) and a few free apps are suggested. Thanks to a helpful tool called CHIRP, programming these radios is simple and feels like filling out a spreadsheet.

Baofeng as an FM Radio

The Baofeng radio's capacity to work as an FM radio for your preferred radio stations is one of its most well-known features. When local radio stations transmit emergency announcements and information during disasters, this feature can be quite helpful.

Just hit the orange call button on the radio's side to switch to FM mode. The SCAN button can then be repeatedly pressed to scan each of the available stations.

The Baofeng radio has memory for up to 128 channels. There is a bright LED light that is included in the radio which is also a very nice feature. It possesses a belt clip but it can also fit nicely into your pocket making it very easy to carry about.

Emergency Frequency

You can choose a frequency by simply entering the required numbers on the keypad, after which you can start transmitting and receiving signals. For example, typing 162.400 will make you able to hear the NOAA weather report. All you have to do is dial 151.940 to reach the most popular national emergency channel.

So that we won't have to try to memorize all of the digits associated with every frequency that we could use, we want to designate specific emergency channels.

To avoid losing frequency while creating a new channel, follow the steps below;

- Tap **VFO/MR** to set the radio into Frequency (VFO) mode.

- Tap **the A/B button** to choose the top frequency. It is worth noting that the arrow to the left of the frequency is shown on the display as it helps to indicate your selection. All programming must be done with the use of the top frequency.

- Switch off TDR/Dual Standby (this should be off by default but ensure you confirm it is). Tap **Menu > 7 > Tap Menu to choose the menu option > With the use of the up and down arrows, choose OFF > Tap Menu for confirmation > Tap the exit key.**

- Using the keypad, enter the frequency that you want to save.

- Press **Menu.**

- Proceed **to option 27.**

- Press Menu again to bring up the channel selection.

- Press the up and down arrows to select the desired channel (000 to 127). It is advised that you start at channel 1 and work your way up to channel 2. A channel that bears "CH" in front of its number indicates that its frequency has previously been preserved.

- To save the frequency to the selected channel, tap **Menu.**

- Lastly, give the **Exit button a tap.**

Activity

1. Explain how Baofeng can be used as an FM radio.

CHAPTER 5
UNDERSTANDING VHF & UHF

One of the most difficult choices to make when choosing the perfect radio is deciding which bandwidth to utilize. The two main wavelengths used in communications between vehicles and between vehicles and bases (Ultra High Frequency) are Very High Frequency (VHF) and Ultra High Frequency (UHF). The last thing you want to worry about in an emergency requiring quick communication is your radio's range. Even if you would not think much of the differences between these bands, this is not the case.

One of the primary distinctions between the two bands will be frequency. Since light travels at a consistent speed, radio waves move all at once since they are a type of light. This is important because the various radio waves are assessed against their "waves" rather than their speed, which cannot be defined. Consider how waves on a body of water resemble radio waves. The number of waves that pass a specific point in a second is known as the "frequency" (measured in Hertz), and the distance between the wave crests is known as the "wavelength." Something's "wavelength" must be shorter for something with a higher frequency than for something with a lower frequency. The reason for this is that waves of a smaller size can only accommodate more waves in a given length of time since light travels at a constant speed.

VHF

VHF operates between 30 MHz and 300 MHz in frequency range. The wavelengths of VHF are occasionally 10 times longer than those of UHF, which has a frequency range of 300 MHz to 3000 MHz. This suggests that because VHF waves are longer, they are harder to break and hence go farther.

Think about how a big wave would affect the rock differently from a little one. A smaller amount of water can flow around the rock because the smaller wave is greatly deflected away from it. The bigger wave can still pass over or around the rock, though. Because UHF radio waves have a higher frequency, they travel farther and fade more easily, making them shorter.

UHF

You may be asking yourself, "Why would anyone use UHF at all?" When in a location with lots of little obstacles, like a stadium race where lots of people are present, you will need a signal that can pass through the gaps in those barriers to connect with the person you are communicating with. UHF signals can go farther than other kinds of signals because of their smaller waves. In scenarios where massive stones are present, like during off-road activities or racing, you will require a larger wave to let the vehicle maneuver around the obstacles. The best alternative for you will be to use VHF because of its much larger wavelength, as UHF waves are shorter and readily deflected by the stones. This is also true for more substantial impediments, like hills or ravines, where the VHF's longer wavelength allows it to pass through the obstructions. The main issue is that the quality will suffer significantly because there won't be a clear line of sight. Because of the greater distance that VHF signals may travel, off-roaders almost always choose to communicate using the VHF band.

Which one is better between VHF and UHF?

The only factor influencing your decision between a VHF and a UHF frequency is the intended purpose. When you're outside and there are no barriers of any kind, you utilize VHF frequently. If there is no obstruction of any kind, VHF frequencies frequently go farther. The only circumstance in which you might need to utilize the VHF is when you are outside in an open area, such as a field. Because VHF radios use much smaller frequencies than other radios, there may occasionally be random interference while using them.

Conversely, UHF is seen to be an all-around superior signal for communication across greater distances. When using radios indoors, such as in buildings or urban areas, UHF is a far better option. You are less likely to be deduced by the other two radios in addition to using UHF. Because UHF signals penetrate concrete, steel, and wood far more effectively than VHF signals can, they are far more effective for usage indoors and can penetrate deeper into buildings.

Who Uses UHF and VHF?

Police, fire, and emergency medical services workers are among the public safety officers who typically use UHF, which has television channels ranging from 77 to 80. Everyday devices like telephones, televisions, and even ham radio operators use UHF. UHF radios are utilized in a variety of environments, such as casinos, security agencies, warehouses, construction, manufacturing, and healthcare, to communicate with people inside the building and between departments. While a ham radio operator on 13 cm uses MHz frequencies between 2300 and 1310, public safety officials use frequencies between 849 and 869.

VHF is a type of radio frequency that's often used for intercom among marine workers and aboard boats (Woodward). This is an extremely vital tool to have on board since it will allow you to speak with other boaters in the region in case specific circumstances arise. Channel 16 is utilized for emergency calls, and there are specific protocols that must be adhered to. Organizations like the California Department of Forestry and Fire Protection (CAL FIRE) and the Transportation Security Administration (TSA) use two-way radio communications via VHF.

How Can Signal Strength be Improved on VHF-UHF Two-way Radio?

One method of increasing a two-way radio's transmission range is to improve the antenna. The length of your antenna directly relates to the length of the radio waves you receive. UHF two-way radio antennas often have an extended and compact shape because UHF (ultra-high frequency) wavelengths are quite short (Personal Radio Services). A somewhat larger antenna is needed to boost a VHF signal's range and potential travel distance. Channels 2 through 13 can be received by antennas tuned to the VHF band; channels 14 through 83 can be received by antennas tuned to the UHF spectrum. (Extremely High Rate).

Given that other frequencies frequently interfere with VHF, determining the precise location of the interference source is the best method to ensure you are not interrupted. For example, there are a lot of potential sources of interference on a boat. Make sure you pay attention to what the recipient is saying and notice if the volume changes.

Another way in which you can help with interruptions is by bonding. Bonding ensures that the noise goes away to the ground instead of it being radiated. All motors and such ought to be constructed in the ground.

One reason why broadcasts can go awry is the frequency overlap problem. This means that if two radios are using the same frequency, the radio waves will interfere with one another and cause the signals to overlap. This is most likely going to happen when they are in the same coverage area or within each other's range.

There won't be any issues if you use a single transmitter, but using multiple transmitters to cover a large region will make the operation more difficult. The reason for this is that you don't want any interference between the transmitters.

Activity

1. Differentiate between VHF and UHF.

2. Improve the signal strength on your radio.

CHAPTER 6
POWER AND YOUR RADIO

Depending on whatever power level offers the optimum range for your particular application, you can broadcast on the BaoFeng BF-transmitter, F8HPs at 1 watt (low power), 4 watts (medium power), or 8 watts (high power). The adjustable power helps when battery conservation is necessary, but if you want the radio to run as efficiently as possible, you can set it to the maximum power. It seems to make sense that as the power is increased to its maximum, the battery life will decrease.

The receiver's performance is the best when evaluated based on all parameters. Even while the selectivity is still a little depressing, this is only apparent in areas with heavy RF activity, such as cities. The filtering issue has little bearing on your experience when you are outside, say, camping or hiking.

Charging your Baofeng Radio

More devices that tend to make life easier are being manufactured in this day and age thanks to technological advancements. This has also led to the emergence of new power sources, or more accurately, a diversity of methods for converting energy into electricity for all of our devices, including the incredible Baofeng radio.

There is more to either cable than meets the eye. You are undoubtedly aware that you can use the BT1013 to connect your BL-5L 3800mah battery to USB or that the 10V USB Smart Cable may be used to convert your radio charger to USB. You can connect your radio charger to USB using one of the two cables. These days, USB ports are available almost everywhere—in our homes, cars, and pretty much everywhere in between.

You can still power and charge your radios while driving even if your car does not have a built-in USB port by using a USB adapter that you insert into the cigarette lighter socket. USB ports are getting more and more prevalent in cars these days, but if yours isn't one, you still have options. How about taking your radios on an excursion where AC or DC power is unavailable? Many solar-powered USB kits are available, so you may charge your radio for any kind of trip— be it an overnight camping trip, an easy stroll through the hills, or an excursion far from civilization.

Follow the steps below to charge your Baofeng radio properly;

- Attach the power adaptor's DC connector to the charger's base.
- Connect the power adaptor's AC connector to a main AC wall outlet.
- Place the radio inside the base that charges it.
- Make sure that the radio and the attached charger are making enough contact. The red LED will turn on steadily when your radio is charging.
- The radio is fully charged when the green status LED on the charger remains constant. Make sure you turn off the radio at the appropriate moment to prevent overcharging, which could harm the battery or the radio itself.

Caution

Battery

Ensure that the radio is well powered off before attaching or removing the battery; you might also decide to rotate the power/volume knob all the way counter-clockwise to ensure it has been turned off.

Installation

Gently press the battery in parallel with the radio's body so that its lower edge is 1-2 cm below the radio's edge. Slide the battery upward until you hear a click, which indicates that it has been securely locked in place, as soon as you see alignment with the guide rails.

Removal

If you would like to remove the battery, tap the Push buckle on the middle top as you continue sliding the battery downwards.

Charger LED Codes

Red LED	Green LED	Status
Flashing	Steady	Standby (charger empty)
Steady	Off	Error (charger with radio) Charging
Off	Steady	Charge complete

It should be noted that the battery and charger have matching notches to allow for proper individual charging of the battery. If you have two batteries, you may charge one and use the other in your radio. This is a highly practical solution. You won't ever miss something significant in this way.

Battery Maintenance

The battery is often provided without power from the factory hence you should ensure the battery is well-charged for at least four or five hours before you commence using your radio.

Follow the tips to ensure that the life of your battery is adequately prolonged;

- Ensure batteries are charged at room temperature only.

- Make careful to turn off the radio when charging a battery that is linked to it to facilitate a quicker charge time for the battery. Make sure you don't disconnect the battery or plug in the charger before the charging is finished.

- Make sure you never charge a wet battery. Before attempting to charge your battery, quickly dry it out if it seems low and wet.

- It is normal for your battery to wear out over time most often for a long period due to Baofengs durability. When the radio has been operated for a very long time ensure you take into consideration the purchase of a new battery so you can replace the one in use before it gets damaged.

- Below-freezing conditions will cause a battery's performance to decrease. It's generally advisable to have a backup battery ready for any task that requires you to operate in a relatively chilly environment. This should ideally be placed inside your jacket or somewhere that ensures the battery will remain warm.

- The connection between your radio and the battery may become impeded by dust. To guarantee correct contact between the radio and charger, clean the contacts if necessary with a clean cloth.

Activity

1. Charge your Baofeng radio.
2. List the steps to perfectly maintain your battery.

CHAPTER 7
ANTENNAS

An antenna is a particular type of transducer that can change electromagnetic (EM) waves into electric current or convert electric current into EM waves. Non-ionizing electromagnetic fields, such as radio waves, microwaves, infrared radiation (IR), and visible light, can be sent and received with the assistance of antennas.

Both in our daily lives and in the majority of commercial sectors, radio wave and microwave antennas are widely used. Antennas for visible and infrared light are utilized much less frequently. They are utilized in a wide range of contexts, even though their uses have become more specialized. The antenna is one of the components of a radio that is thought to be extremely crucial. The most important aspect influencing the overall quality of a transceiver's signal is its choice of antenna. This is accurate although battery life, power, and radio type are all important factors.

An appropriately tuned antenna is necessary to maximize the performance of any radio. The market is today filled with a large variety of antennas, and choosing the correct one can affect the radio's total lifespan in addition to its performance. Still, there are some limitations even in the face of a wide range of options. Base stations and mobile radios provide you with a lot of flexibility when choosing an antenna, but walkie-talkies and handheld transceivers pose other kinds of difficulties.

A handheld two-way radio may feature either a fixed or removable antenna, depending on the manufacturer's preference. The housing of the radio itself incorporates a fixed antenna, which cannot be detached for any reason. Consumer FRS and GMRS radios, as well as some low-wattage commercial radios like the Motorola CLS1410 and RMU2080d, both feature antennas that are permanently attached and secured in place.

Most high-end, high-wattage handheld two-way radios intended for business usage have detachable antennas that are specifically tuned to the bands and frequencies that each model uses. This means that, theoretically, a business radio operating on 403–470 MHz frequencies and a UHF antenna set to those frequencies should work together. There is a catch, though. There is no assurance that an antenna of this type can be swapped out for another, even if it can be disconnected. This is because there's a chance that the connectors on the radio and the antenna are different.

How Are Antennas Categorized?

Antennas are frequently divided into transmitting and receiving categories. However, a transceiver included in many antennas enables them to perform both tasks. A gadget that transmits also provides a current to a transmitting antenna. The antenna generates electromagnetic waves at a specific frequency as a result of this current. These waves then radiate through the atmosphere and can be picked up by one or more other antennas.

An FM signal is a particular type of radio wave that is used by radio stations to deliver music within the electromagnetic spectrum. The station's transmitter sends the electric current carrying the music to the antenna at the desired frequency. The antenna transforms the electric current into radio waves, which are subsequently directed in every direction. A receiving antenna receives electromagnetic waves that are transmitted through the atmosphere. A small amount of current is generated by the antenna as a result of these waves; the quantity of current generated is proportional to the strength of the signal. The current is then sent to the receiving device, which modifies it so that it can function properly in the new environment.

For instance, the antenna on a car could be able to pick up the FM signal broadcast by the radio station. The radio waves of the transmission are changed into current by the antenna, which is then supplied to the radio in the vehicle. The radio both boosts the current and alters it in other ways before transmitting the altered signal to the speakers in the form of music.

Antennas are primarily used for electromagnetic wave transmission and reception at predefined frequencies. Their design is also influenced by other factors, like the signal's strength and direction of travel. This explains the large range of shapes, sizes, and forms that antennas come in. Television antennas, for example, are built quite differently from car antennas, and both of these kinds of antennas are built very differently from cell phone and microwave antennas. There are numerous other varieties of antennas with extremely distinct constructions.

Main Types of Antennas

Depending on the design of the antenna, different use cases may be supported by the device. It is common practice to classify them into several categories to facilitate easier differentiation between the various kinds, although there is no consensus throughout the industry regarding the components that make up each category. Despite this, there are a few standard categories that are frequently used when describing and differentiating one type of antenna from another.

- **Aperture**: An antenna that has an opening in its surface that helps with direct EM transmission or reception to get a much larger gain. The size and shape of the antenna are based on how the antenna is being used. Aperture antennas are often deployed in circumstances that need flush mounting, like that of an aircraft or spacecraft.

- **Array**: a single radiation pattern is produced by an antenna composed of smaller connected antennas that sync together. Array antennas provide far more control over their directionality and can decrease interference as well as boost gain. Array antennas are used in many different contexts, such as military radar systems, 5G networks, and wireless communications.

- **Reflector**: an antenna that has one or more components that show the EM waves to better focus or direct them. Reflector antennas are most often utilized in microwave and satellite communications. Many have a parabolic structure that shows EM waves like those used in satellite dishes.

- **Lens**: an antenna with a glass, metal, or dielectric material embedded lens. Often at a very high frequency, the antenna transmits or receives electromagnetic waves by taking advantage of the convergence and divergence characteristics of the lens. Microwave communications and radar systems both frequently use lens antennae.

- **Log periodic**: This directional antenna supports a wide range of frequencies thanks to its diverse elemental design. The size and arrangement of the elements determine the supported range, which in turn depends on a logarithmic function of frequency. When a circumstance calls for changeable bandwidth or support for high-frequency communications, such as analog television, cellular communications, or shortwave radios like the Baofeng radio, log periodic antennas can be quite helpful.

- **Microstrip**: this is a small antenna printed into a circuit board. The actual antenna is a patch made of conductive material that is put on a dielectric substrate and rests on a ground plate by itself. Cell phones and other wireless communication and mobile devices make heavy use of microstrip antennas.

- **Traveling wave**: The electromagnetic waves passing through a directional antenna only travel in one direction. On the other hand, waves traveling in several directions are characteristic of most other types of antennas. The waves can tolerate a wider range of frequencies since they only go in one direction. Traveling-wave antennas are used in a wide range of applications, such as amateur radios, telecoms, and analog televisions.

- **Wire**: an antenna that is connected to a transmitter or receiver at one end and is made out of just one wire strand. The most portable and easiest to build antennas are those composed of wire. They are heavily utilized by radios, cars, buildings, ships, airplanes, and a host of other devices and constructions.

Positioning the Baofeng Antenna

The Baofeng radio has two major types of radio which are the Male SMA connector (transceiver) and the Female SMA connector. Get the two connectors well aligned and then turn in a clockwise manner until it stops.

Follow the set of rules to ensure you get your antenna properly fixed;

- To prevent harming the antenna, the outside materials, and the connecting base, make careful not to over-tighten it.

- Make sure you hold the antenna by the base screw when inserting it.

- To protect the transceiver from harm, make sure the external antenna you choose to use has an SWR of at least 1.5:1.

- To prevent any kind of interference to the transceiver, wrap the outside of the antenna or refrain from handling it in your hands.

- Make sure you always utilize an antenna when transmitting.

Activity

1. What are antennas and how are they categorized?

2. Briefly describe the types of antennae.

3. How should the Baofeng antenna be positioned? Position your antenna properly.

CHAPTER 8

WHAT BAND/FREQUENCY DO I USE?

The frequency of a repeating event is the number of times it occurs within a given period. Occasionally, it is also called temporal frequency for clarity's sake. There is a difference between this frequency and angular frequency. The hertz (Hz), or one occurrence every second, is the unit of measurement for frequency. The period can be thought of as the opposite of frequency because it is the length of time that passes between events.

A heart's period, T, or the interval between beats, is equal to 60 seconds divided by 120 beats, or 120 beats per minute, or 2 hertz. This is an example of a heart beating at this frequency. Frequency is an essential property that is used to describe the pace of oscillatory and vibratory events in the domains of science and engineering. Radio waves, light, mechanical vibrations, and audio transmissions are a few instances of this kind of event.

Focusing only on radio, the term "radiofrequency," or "RF," describes the rate of oscillation of a mechanical system or a magnetic, electric, or electromagnetic field in the frequency range of around 20 kHz to 300 GHz. The oscillation rate of an alternating electric current or voltage is another term for radiofrequency. These are the frequencies that fall roughly between the upper and lower bounds of audio and infrared frequencies, at which energy from an oscillating current can radiate off of a conductor into space as radio waves. Depending on the source, there are several different ways to specify upper and lower bounds for the frequency range.

FCC RULES

The first piece of information you need to know is the frequency ranges that have been approved by the FCC for a given license class. The frequency privileges that an operator is permitted in the HF bands are largely determined by the license class they possess. All license levels—Technician and above—have access to the same frequency allocations above 50 MHz. In particular, the frequency range of the 2m band is 144 MHz to 148 MHz. The FCC Rules state that the band with a frequency range of 144.100 MHz to 148.000 MHz is open to the use of any mode, including FM, AM, SSB, and CW, among others. The FCC has mandated that only CW transmissions can take place in the frequency range of 144.0 to 144.100 MHz.

Band Plans

Understanding FCC frequency authorizations can be an excellent place to start. Different modulation techniques are used by amateur radio operators to communicate. Since a radio set to one modulation type cannot receive a signal of another, these modulation schemes are typically incompatible. For example, an FM receiver cannot receive an SSB signal, and vice versa. It is vital to share the band with other users and avoid undue interference to make sensible use of the allocated frequencies. Therefore, having a band plan that splits the band into sections for each type of operation makes perfect sense.

The 2-meter frequency amateur band plan of the ARRL allows a wide variety of radio operations. A sizable chunk of the band has been reserved especially for FM transmissions due to the widespread use of the FM mode. The band has been divided into sections designated for repeater inputs and repeater outputs. The output frequency of the repeater is the frequency that users adjust to receive it (which is the frequency individuals broadcast to use the repeater). Note that the frequency at which these segments have been spaced is 600 kHz, which is the typical offset for 2-meter repeaters. Furthermore, certain frequencies are assigned specifically for FM simplex broadcasting.

The most popular kind of VHF radios are basic FM mobile or portable transceivers. Usually, these radios tune the whole 2 MHz range, from 144 MHz to 148 MHz, in 5 kHz steps. The right frequency range for FM operation is laid forth in the band plan, but the story goes deeper than that. FM operation is "channelized," meaning that certain 2m FM frequencies are designated under the band plan. Repeaters require careful channel management because they move in frequency slowly and need to be synchronized to reduce interference. All stations are supposed to operate on frequencies that are spaced just enough apart to provide the transmission without interfering with nearby channels. You might expect the channel spacing to be 5 kHz, which is the tuning step of most FM radios. This does not work because a typical FM signal has a bandwidth of about 16 kHz.

When using a repeater, simply enter the published repeater frequency and set the transmit offset, which is usually +600 kHz or -600 kHz for a 2-meter band repeater. Non-standard repeater offsets may be used in some parts of North America, as indicated in the repeater directory. You must set the proper tone frequency on transmit for repeaters that require a CTCSS tone for repeater access.

Selecting the appropriate simplex frequency might be challenging since it depends on whether the channel spacing in your area is 15 kHz or 20 kHz. In all of North America, the National Simplex Frequency—also referred to as the calling frequency—is 146.52 MHz. In regions that employ 15 kHz channels, the neighboring channels, going upward, are 146.535, 146.550, 146.565 MHz, and so on. Below the calling frequency are the following: 146.505, 146.490, 146.475, and so forth. Areas that employ 20 kHz channels have the following frequencies: 146.500, 146.480, and 146.460 MHz moving down, and 146.540, 146.560, and 146.580 MHz moving up.

Oftentimes there is always another group of FM simplex frequencies in the 147 MHz. The typical layout of simplex channels can be found below. Nevertheless, you need to take into consideration that your local band plan might be different than this.

2m FM Simplex Frequencies (typical usage, check your local band plan)
15 kHz Channels

146.400, 146.415, 146.430, 146.445, 146.460, 146.475, 146.490, 146.505, 146.520, 146.535, 146.550, 146.565, 146.580, 146.595, 147.405, 147.420, 147.435, 147.450, 147.465, 147.480, 147.495, 147.510, 147.525, 147.540, 147.555, 147.570, 147.585

20 kHz Channels

146.400, 146.420, 146.440, 146.460, 146.480, 146.500, 146.520, 146.540, 146.560, 146.580, 146.600, 147.400, 147.420, 147.440, 147.460, 147.480, 147.500, 147.520, 147.540, 147.560, 147.580

160 meters (Top Band) 1.800 -2.00 MHz

The top band is the lowest frequency ham radio allocation. Although it is known to be one of the short wave bands and it is most times mentioned with the other HF amateur radio bands, to be quite exact, it is basically in the MF portion of the spectrum.

Top Band is however not allocated for ham radio use in all countries and the exact limits of the bands might vary. Generally, the maximum extent for any allocation ranges between 1.8 and 2.0 MHz.

The Top Band is an allotted amateur radio frequency range that resembles the Medium Wave broadcast band in many ways and overlaps with it. Ham radio contacts are usually maintained on a more local level throughout the day when signals can be received via ground wave and the strength of the transmitter and antenna. Radio signals may be received from stations hundreds of miles away during the night when the ionosphere's D layer is at its thinnest and can travel farther.

If stations in North America and Europe have high-quality antennas on both ends, they can make transatlantic contacts when propagation conditions are favorable. Longer distance connections are also possible. When making lengthy contacts on Top Band, the entire path should be dark. However, for encounters with people on the other side of the world, there may be notable enhancements during the day. The longest these improvements could endure is ten to fifteen minutes, maybe even less.

When traversing shorter distances, like the one connecting North America and Europe, signal strength is strongest at either dawn or nightfall on both ends of the journey. Long-distance, north-to-south routes usually have their peak travel times around midnight.

Long-distance work is known to perform better in the winter months due to the increased number of hours spent in darkness and the decreased amount of static. As a result of the fact that this does not coincide with the optimal conditions in the other hemisphere, it indicates that these signals may be received at any time during the year.

80 meters 3.500 - 4.000MHz

Ham radio in either the 80- or 75-meter bands is under the high-frequency (HF) category of the electromagnetic spectrum. The precise distribution is determined by the radio area in which the country is situated. In most cases, this can be anywhere between 3.5 and 3.8 MHz; however, in North America, frequencies up to 4.0 MHz can be used; nonetheless, there is a broadcast band allocation above 3.8 MHz.

The ham radio band is shared with other services, therefore it is frequently busy, which can lead to a lot of background noise, especially at night. There is also a chance of extremely high static levels.

During daylight hours, stations can be heard up to around 100 miles distant; this makes it a perfect band for medium-range communications. Further-off ham radio stations can be heard at night. Over a thousand miles is usually the norm, and radios with excellent antennas may receive signals over even longer distances. The band can succeed at any time, but it shines during the sunspot minimum years.

The line where dawn and dusk fall is called the gray line, and it can produce very high reception, with stations from the other side of the world being audible at the same strengths as many local stations. However, it's possible that this won't continue long and that it may only happen in certain places. Furthermore, spring and fall are the best seasons.

Most ham radio SSB DX activity is concentrated in the "DX window" in the top 25 kHz of the European band. This means that there should never be any obstacles in the way of this section of the band. This should be investigated even if it seems unlikely that any DX stations will be able to receive transmissions. This is because stations with good antennas and positions may be able to pick up DX signals.

Ham radio stations in North America and certain other areas of the world have an allocation up to 4.0MHz hence it is common to work split frequency with stations that do not have this allocation, making use of the DX window beneath 3.8MHz for European stations and above 3.8MHz for North America, etc.

40 meters 7.000 - 7.300MHz

An especially useful amateur radio band is 40 meters, which provides an interesting combination of global communications at night and short-haul DX during the day. Even though the band has been widened to 200 kHz throughout Europe, there can still be some broadcast stations operating in the 7.100–7200 MHz area. In North America, where frequencies as high as 7.3 MHz are still available, interference from European broadcast stations, who are assigned duty for this portion of the spectrum, can present difficulties.

During the day, it is not unusual to be able to hear radio stations that are hundreds of miles away. Then, as night falls, the intensity of locally transmitted stations declines, and the range over which stations may be heard greatly increases. This band is preferred by many during the sunspot cycle phase when there are fewer sunspots because long-distance contacts can be made during the night. Again, there is a chance that the gray area will yield surprising results.

This channel might be a good hunting area for ham radio operators with medium-power transmitters and moderate-sized antennas. The usage of directional antennas by radio hams is quite uncommon; as a result, average amateur radio stations are less at a competitive disadvantage. When used in conjunction with a reliable earth or ground-plane system, trap verticals can perform well. This makes it possible to contact and connect with stations anywhere in the world.

30 meters 10.100 - 10.150

This band specifically was released for amateur radio use after the World Administrative Radio Conference which was held in 1979. Although it has been around for so many years, it is still not widely used even though it can provide very good results.

Due to its similarities to 40 meters, this ham radio frequency can be used for DX contacts during the majority of the day. However, it is generally better at night, allowing for international ham radio interactions. It has also been found that absorptions are often low during sunspot minimum periods, when ionization levels are lower, to allow for long-distance connections throughout the day.

This frequency, together with the 40-meter ham radio band, is useful for DXers without particularly strong stations, as are the other WARC bands. Not many typical directional Yagi antennas have this frequency, and some stations may still be using linear amplifiers, which are not suitable for this region. Consequently, it suggests that the operational disadvantage for those with stations closer to average will be lessened.

The majority of the operation is carried out in Morse due to the limited space available on the band, which is also shared with other services. This results in a high level of commercial activity. The IARU for Region 1 has suggested that phone operations and contests be prohibited on the frequency.

20 meters 14.000 - 14.350 MHz

The main long-distance band for radio amateurs is this amateur radio allocation, which offers consistent long-distance communication throughout the sunspot cycle. There are essentially no limits on the locations where ham radio activity is allowed, and the band allotment is the same everywhere in the world.

Stations between 500 and 1500 miles away are nearly usually audible during the day, and under favorable conditions, stations up to roughly 2000 or 3000 miles away can be heard. Often at night, the band shuts, especially in the winter and around sunspot minimum. Good results are usually obtained in the spring and autumn, with stations in the other hemisphere audible at different times of the day.

All across the world, signals are audible throughout the day. Early in the morning, transmissions from the east arrive; these signals are usually from across the globe. As these signals diminish, new local signs will emerge, and as the Sun rises in that direction, openings might appear to the west. Opportunities to the west can arise as the afternoon wears on. There might be more chances to visit places across the globe as their morning draws near. Long-distance stations will be to the west of local signals as they get weaker in the evening due to a decrease in ionization levels.

As the primary ham radio DX band, 20m is frequently crowded, and when any rare amateur radio stations appear, competition is fierce. As a result, many ham radio stations that operate in this band employ good directional antennas mounted high up, in conjunction with high transmitter powers. Some of the "big" stations use kilowatt power (where licensing conditions allow) and at least three-element Yagi antennas at a height of around 60 feet (20m). It is still feasible to establish a lot of beneficial contacts, but effective working methods are necessary. In favorable conditions, it might be important to evaluate any pile-ups heard and determine whether to stay and establish contact with a particular station or to move on and see whether contact is more likely with other DX stations.

17 meters 18.068 -18.168MHz
This was made available for amateur radio following the WARC 79 conference, just as the 30m band. Reviews suggest that some older transceivers might not be able to handle this ham radio allocation. It is very much a midway house between 15 and 20 meters in terms of performance. Even if it is rather limited, it is nevertheless fairly well-liked and worthwhile to check into when the circumstances seem pretty favorable.

Radio amateurs with more common stations may find great opportunities to contact uncommon DX stations on this amateur radio channel. The majority of stations still utilize dipoles even though beam antennas for the band are available. This is because the few powerful stations that have beams may employ them for the more conventional DX bands of 10, 15, and 20m. On the other hand, an increase in WARC band antennas indicates an increase in the number of users of these frequencies.

15 meters 21.000 - 21.450
The state of the sunspot cycle has a greater impact on the conditions encountered in this amateur band than it does on the 20-meter band. When it is at its busiest, it is accessible day and night and can assist thousands of miles of propagation. Usually, things aren't the best in the morning but become better during the day. There may be few stations audible during the day and none audible at night during the solar minimum. The 13-meter broadcast band is located at the top of the 15-meter ham radio band. It is entirely possible to have this observed to quickly determine whether the amateur band may be open.

12 meters 24.890 - 24.990
This amateur radio band is the highest of the ham radio bands available at WARC 79. As a result, it is not as widely used as the traditional bands such as 20 meters, 15 meters, and 10 meters, but it can still provide some good results and has a reasonable level of occupancy when compared to 15 or 10 meters. This band, like 17m, is quite narrow, but it is worth investigating when conditions indicate that the band may be open. Furthermore, there are few stations using beam antennas, making it a good hunting ground.

10 meter 28.000 - 29.700
In the short-wave range of the spectrum, this is the highest known frequency for the amateur radio band. This allotment is fixed globally, and given its bandwidth (1.7 MHz), it is utilized for a variety of transmission modes, such as FM, SSB, and Morse. Additionally, several countries have ham radio repeaters that, under the right circumstances, can provide global coverage. Regarding its characteristics, it may only enable ionospheric propagation through occasional E, which primarily occurs in the summer, at the sunspot minimum. This allows transmission across a distance of roughly one thousand kilometers.

When the sunspot cycle reaches its zenith, it presents ideal conditions for long-distance communication and emits incredibly powerful signals. It's also commonly known that this frequency makes it possible for ham radio stations with weak antennas and modest watts to establish contacts across longer distances. In general, for frequencies like this to propagate, the signal route must be in daylight. Most of the time, activity in the SSB band is concentrated between the beacon region and 28.60MHz, and sometimes much higher. However, since certain stations may also be involved in this industry, it is worthwhile to look above this, particularly in terms of content.

Activity

1. What are the frequencies that can be used on the Baofeng radio?

CHAPTER 9
TALK ABOUT REPEATERS

In the telecommunications industry, a repeater is a piece of electrical equipment that receives a signal and then retransmits it. To increase the signal's range or enable reception across obstacles, repeaters are utilized to extend transmissions. Multiplexing is the method used to achieve this. Different types of repeaters use different frequencies or baud rates, for example, but they all transmit the same signal in the same way.

An information-carrying signal steadily loses power as it moves across a communication channel, resulting in increasing degradation of the signal's quality. For example, because of the copper wire's resistance, some of the power in the electric current representing the audio signal is lost as heat during a phone conversation across a wired telephone line. This occurs as a result of the current flowing through the copper wire.

Types of Repeaters

Telephone repeater
This type of repeater can be used to increase signals in a telephone line.

Landline repeaters
This type of repeater is commonly used with trunk lines that carry long-distance calls. Multiple pairs of wires make up an analog telephone line. Additionally, this line has a transistor-based amplifier circuit that receives power from a direct current (DC) source. The goal of this circuit is to make the audio signal that goes over the line with alternating current stronger. The wire pair comprises not one, but two unique audio signals, one moving in each direction, because the telephone is a duplex communication device. To avoid creating a feedback loop, telephone repeaters must be bilateral, which means they must magnify the signal in both directions. This requirement makes the design of telephone repeaters highly hard.

One of the first repeaters ever made and one of the earliest uses of amplification technology was the telephone repeater. The development of telephone repeaters allowed people to have long-distance phone conversations between 1900 and 1915. Nowadays, fiber optic cables—which employ optical repeaters—make up the great bulk of the cables used for telecommunication.

Optical Communicator Repeater
The purpose of this kind of repeater is to extend a fiber optic cable's signal range. Like brief light pulses, digital information travels across a fiber optic connection. Photons, the units of light, can either be absorbed or dispersed throughout the fiber. A phototransistor aids in the conversion of light pulses to an electrical signal, an amplifier boosts the power signal, an electronic filter reshapes the pulses, and a laser converts the electrical signal back to light before sending it out the other fiber in an optical communications repeater. However, optical amplifiers are being produced for repeaters to amplify the light itself without there being a need of having to convert it to an electric signal first.

Radio Repeater

A radio receiver and transmitter that retransmits a radio signal is called a radio repeater. An amplifier installed within a phone line is called a telephone repeater. An optoelectronic circuit that strengthens the light beam inside an optical fiber connection is called an optical repeater. Repeaters come in a variety of varieties. Radio and television signals are disseminated via a sort of repeater called a broadcast relay station.

Broadcast relay station, re-broadcaster, or translator

This type of repeater is one used in the extension of the coverage of a radio or television broadcasting station. It has a secondary radio or television transmitter. The signal from the primary transmitter most times comes over leased telephone lines or by the relay of the microwave.

Microwave relay

This is considered a unique point-to-point communication link since it has a microwave transmitter that transmits the information to the next station via a different microwave beam and a microwave receiver that receives information from another relay station over a line-of-sight distance. Over vast stretches of the continent, networks of microwave relay stations transport computer data, television shows, and phone conversations from one city to another.

Passive repeater

This is a microwave relay that consists of nothing more than a smooth metal plate that is used to reflect the microwave beam in a different path. When signal amplification is not required, it can be utilized to transmit microwave relay signals over topographic obstacles such as hills and mountains.

Cellular repeater

This radio repeater can enhance mobile phone reception within a designated area. With a directional antenna, an amplifier, and a local antenna to rebroadcast the signal to nearby mobile phones, the gadget serves as a compact cellular base station. The directional antenna receives the signal from the nearest cell tower. In downtown locations, it is commonly utilized in office buildings.

Amateur radio repeater

Amateur radio operators use this to communicate back and forth over a region that would be challenging to cover with point-to-point techniques on VHF and UHF frequencies. Such repeaters are usually constructed and maintained by clubs or individual operators, and the public is welcome to use them as long as they are licensed amateur radio operators. A hilltop or hillside is the best location for a repeater because it will optimize the device's usability over a large area.

Radio repeaters make the coverage of communications much better in systems that make use of frequencies that typically have line-of-sight propagation. When there is no repeater, these systems are limited in range by the curvature of the Earth as well as the blocking effect of the terrain of very high buildings. A repeater on a hilltop or a tall building can give room for stations that are out of one another line-of-sight range to enable reliable communication.

Radio repeaters can also facilitate translation across different radio frequency sets, allowing two different public service agencies (such as a city's police and fire department or neighboring police departments) to communicate with each other in a coordinated fashion. As an extra detour between the source and the destination, they may additionally offer links to the public switched telephone network or satellite networks (BGAN, INMARSAT, or MSAT).

Repeater stations listen on frequency A and transmit on frequency B, respectively, in most situations. All mobile stations use channel A for transmission, whereas channel B is used for receiving signals. The frequency at which the system is functioning may be much higher than the difference between the two frequencies, even if we take 1% as an example. Duplexers are incredibly selective filters that distinguish between the outgoing transmission signal, which is billions of times stronger than the weak incoming received signal. The repeater station will typically use the same antenna for both reception and transmission. It may be necessary to use separate sending and receiving stations, which are then linked together by either a wireline or a radio link. Mobile units do not need to be fitted with cumbersome and pricey duplexers because they only broadcast or receive at any one moment. This is in contrast to the repeater station, which is designed to facilitate simultaneous reception and transmission.

Using a Repeater

A transceiver that can broadcast on the frequency used as the repeater's input and receive on the frequency used as the repeater's output is required to use a repeater. Depending on the band under consideration, the fixed amount separating the input and output frequencies is used. The offset is the name given to this difference. To provide one particular example, the offset is 1.6 MHz on 1.25 meters. For example, a repeater running on 1.25 meters may have an input frequency of 222.32 MHz and an output frequency of 223.92 MHz. Repeater frequencies are typically described in terms of the output frequency (i.e., the frequency that you set your receiver to listen on) as well as the offset. Your receiver is tuned to a frequency that is offset in frequency by some amount relative to the frequency on which your transmitter runs.

Most transceivers that are intended for use with FM repeaters already have the proper offset pre-configured. They typically have a switch that lets the user select between duplex operation (transmit and receive on distinct frequencies) and simplex operation (transmission and reception on the same frequency). Therefore, in the above example, to utilize the repeater, you would need to switch your transceiver to duplex mode and dial 223.92 to hear the repeater.

When you start transmitting, your radio will automatically change frequencies to the repeater input frequency of 222.32 MHz, which is 1.6 MHz lower in frequency.

Once you have the correct frequency selected, all you need to do is hit the microphone button to transmit through the repeater. You will be able to "access" the repeater as a result Most repeaters are open, meaning that anyone in their range is welcome to use them. However, there's a chance that some repeaters won't be accessible. They are only used by specific groups of people, like members of a specific club. Only the broadcast of a continuous tone below the hearing range or a brief "burst" of tones will allow access to these restricted repeaters.

This class of tones is known as either PL tones (a Motorola trademark) or CTCSS tones (continuous tone-coded squelch system). Additionally, some repeaters are accessible to all users; but, to use them, you must either listen for a tone that is below your hearing range or enter a secret code. For "open" repeaters, access tones are necessary because they shield the repeater from outside transmission interference that could cause someone to key it by accident. For this reason, open repeaters need access tones. If you are interested in joining the group that manages the closed repeater, get in contact with the operator.

Finding a Repeater

Repeaters serve the majority of communities in the United States. While 2 meters has the most repeaters (over 6000), there are also over 1600 repeaters on 222 MHz, over 5000 on 440 MHz, over 70 on 902 MHz, and over 200 on 1270 MHz. More repeaters are being installed all the time. Repeater frequencies are chosen with the help of frequency coordinators, who are individuals or groups who recommend repeater frequencies based on potential interference and other factors.

There are multiple ways to find the nearby repeater (s). Ask the closest radio club or the local amateur radio community. An extensive list of repeaters in the US, Canada, Central and South America, and the Caribbean is published annually in The ARRL Repeater Directory by the ARRL. The Directory helps learn about local repeater activity as well as for locating repeaters to utilize on business and leisure journeys. An additional helpful resource for travelers is the TravelPlus for RepeatersTM CD-ROM.

FM operation is restricted to specific segments of each band. On 1.25 meters, for example, repeater inputs can be found between 222.32 and 223.28 MHz. The corresponding output frequencies are 223.92 to 224.98 MHz.

Operating a Repeater

It is more convenient for you to become familiar with specific repeater operation practices before attempting to make your first FM repeater contact. It's worthwhile to take a few minutes to listen and make sure you become acquainted with the practices followed by some other local hams. The protocols that are approved may vary slightly amongst repeaters.

First Transmission

To send your first message over a repeater, all you have to do is sign your call. Just shout "N1GZO" or "N1GZO listens" to get attention if the repeater is silent. After you finish broadcasting, you should hear the unmodulated repeater carrier for a brief moment. The repeater is functional based on this squelch tail. Someone interested in chatting with you will get in touch with you following your initial transmission. Certain repeaters have guidelines about how to be heard. But for the most part, your call sign is adequate.

Never initiate a discussion on a repeater using CQ. Sending your call sign is quicker than finishing a CQ. (In certain places, it's okay to just have one "CQ"). Efficient communication is the aim. You're not trying to get someone who's leisurely tuning across the band to notice you on a high frequency. Stations in FM mode might choose to monitor their preferred frequency or not. There is minimal tuning over the repeater bands, except for scanner functioning.

To join an ongoing conversation, send your call sign during a transmission break. The next station to transmit will usually acknowledge you. If you want to use the repeater in an emergency, don't use the word "break" to join a conversation. To make a distress call over a repeater, say "break" followed by your call sign to notify all stations to stand by while you handle the emergency.

Another word on emergencies: Normal rules may be suspended in an emergency, according to FCC regulations, regardless of the band, mode, or class of license. When you receive a distress call, try your best to get in touch with the station that needs help and notify the appropriate authorities right away. End your QSO right away and answer the emergency call if you hear someone calling for help while you are speaking with another station.

When not in use, just dial both calls to switch to another station. Say, "N1II, this is N1BKE," for instance. When calling another station, wait until the talk seems to be coming to an end if the repeater is being used. If it seems like the talk is going on forever, just send your call sign-in during the pauses in their messages. After being acknowledged, ask for a brief phone conversation. Usually, the other stations will wait. Keep your call brief. Try meeting on a different repeater or a simplex frequency if your friend answers. If not, ask your friend to wait until the current chat is over.

An autopatch

Repeaters can make phone calls over the repeater by using autopatch. To access and dial through the system, you must generate the normal telephone company tones if you wish to utilize the majority of the repeater autopatches. Essentially, the transceiver is attached to a telephone-style tone pad that is used to generate the tones. Manufacturers of equipment offer tone pads as standard or optional accessories. Usually, they are attached to the rear of a stationary or mobile transceiver microphone or the front of a portable transceiver. The same autopatch operating procedures apply to whatever devices you utilize.

There are strict guidelines for using autopatch. The first question you should ask is, "Does the call need to be made?" There is no problem if it is an emergency; just do it! It is acceptable to request an ambulance or a tow truck. The remaining six reasons may be ambiguous. As a result, some repeater groups expressly prohibit the use of autopatch except in emergencies.

When regular phone service is available, do not use an autopatch. Most evenings in any metropolitan area, one example of poor operating practice can be heard. Someone will call home to notify you of your departure from the office. Why not call from work before you leave?

Steer clear of using the autopatch for any kind of professional correspondence. You are not permitted to use amateur radio for your employer or to conduct business, according to the FCC. On the other hand, you can communicate personally using amateur radio. You are no longer prohibited by the rules from using the autopatch to place an order for takeout or to make an appointment with your dentist or doctor. To avoid making a toll call, only use an autopatch. The right to autopatch is granted by the FCC. Autopatch capabilities can be abused, and then everyone could lose them.

Activity

1. What are repeaters?
2. Mention two (2) types of repeaters.
3. How can a repeater be found?
4. What is an autopatch?

CHAPTER 10
RADIO NETS AND FINDING THEM

A "radio net" is a network of three or more radio stations that share a frequency or channel for mutual communication. A net is simply a two-way radio conference call that is moderated and conducted under what are known as half-duplex operational conditions. A predetermined set of operating rules must be scrupulously followed during a half-duplex operation to prevent inefficiencies and preserve some sense of order.

The nets can run according to a timetable or continually (continuous watch). Only at specific, established hours and in compliance with a predetermined intercommunication schedule is traffic handled on schedule-operated networks. Continuously functioning networks are always ready to handle incoming traffic since operators are on duty at every network station. This ensures that no traffic is ever lost. When possible, signals relevant to scheduling will be conveyed by a signal communication mechanism other than the radio.

The proper operation of the radio net is frequently the responsibility of a manager, sometimes known as the Net manager. He is a network manager who oversees the setup and continuous operation of multiple sessions of a network. This individual will be in charge of selecting the participants, the format, the day and time, and the net control script. Furthermore, the net manager selects the Net Control Station for each net; in certain cases, especially in smaller businesses, they may even personally handle that duty.

Radio nets have some basic operations you should know and they are as follows;

- Enable participants to conduct ordered conferences among participants that often have common information needs or similar duties to conduct.

- Are characterized by strict adherence to standard formats and procedures, and

- Are usually always responsive to a common supervisory station known as the "net control station" which allows access to the net and also maintains net operational discipline.

The basic Structure of the Net

Nets are characterized by an opening and a closing, with the roll call usually taking place following the opening. This is followed by standard net activity, which could include notifications, official business, and message passing. Military nets will always have the essential opening, roll call, late check-ins, and closure procedures, even with a severely simplified and opaque version of the structure outlined below.

The main principle behind a net should be the same as that of the inverted pyramid used in journalism: the most important messages should come first, followed by ever-lower importance material.

- **Net opening**
    - Identification of the NCS.
    - Announcement of the normal date, time, and frequency of the net.
    - Purpose of the net.
- **Roll call**
    - A call for stations to check in, most times is from a roster of normal stations.
    - A call for late check-ins (stations on the roster who probably didn't answer to the first check-in period).
    - A call for guest stations to get checked in.
- **Net business**
- **Optional conversion to a free net**
- **Net closing**

Every network has a specific goal, and since this usually happens during the network business phase, it is usually not at odds with the organization that manages the network. Usually, amateur radio nets are simply utilized to give stations the chance to trade certain equipment or discuss their most recent operational activity. Transmitting official messages, or radiograms, is the main purpose of net activity for Military Auxiliary Radio Systems and National Traffic System nets.

Modes of Net Operation

Directed Net

This is a net in which no station other than the net control station has the power to communicate with any station except for the transmission of a very important and urgent message, without having to get the permission of the net control station first.

Free net

This is a net in which just about any station can communicate with any other station in the same net without the need to first get permission from the net control station to do so.

Amateur Radio Net

A group of amateur radio operators that communicate with one another "on the air" is called a "ham net" or "amateur radio net." The majority of radio communication groups, often called "nets," get together regularly at a set time and frequency. This is done for a variety of reasons, including emergency circumstances, managing severe weather (such as during a Skywarn activation), group discussion topics of interest, and message delivery.

Amateur Net Operation

Nets operate at different levels of formality depending on how they are organized and what their objectives are. When disaster strikes, a network of nets might come together and work together to achieve a common goal, such as sending out emergency signals. One such network is the National Traffic System (NTS), which is managed and coordinated on a local and national level by members of the American Radio Relay League (ARRL) to handle both routine and emergency messages.

Formal Operation

There is only one net control station (NCS) that oversees the functioning of a formal or directed net during a given session. The NCS operator starts the net at a certain time, calls for participants to join periodically, waits patiently for them to respond or check-in, monitors the roster stations for that particular session, and then generally coordinates the net's operations.

Then, it becomes challenging for a single NCS to maintain control over a net spanning a very large geographical area, such as a continent or perhaps the entire globe. To manage a large region, a net needs to operate on a frequency where signals can travel very far. Oddly, the same ability to transmit over vast distances also leads to a situation where stations that are too close to one another are unable to communicate. In this case, real-time communication between two or more geographically dispersed NCSs can let them stay in contact with all participants who are needed.

A tactical net is a kind of directed net in which stations are assigned tactical call signs to aid in the efficient handling of messages. These nets are typically governed by stricter regulations than conventional radio nets. Using tactical designations like Medical One or Incident Command, participant stations are usually permitted to refer to one another in a tactical net. This makes it possible for the caller to avoid making mistakes or forgetting legal call signs, which can obstruct net progress. Tactical call signs are in no way superior to legal call signs, which the participating stations must still proclaim at the appropriate moments.

Informal Operation

An informal net may or may not have a net control station, but it will not follow the same formalities and protocols as on-air non-net activities. This protocol and its formality would differ from those used in off-net on-air events. As an alternative, it might begin ad hoc, with whoever turns up first, regardless of participation rank, at the appointed time and frequency. In club nets, such as those used to debate various equipment or other subjects, an NCS is merely employed to control the order in which players broadcast their opinions to the group using a round-robin technique.

Types of Amateur Nets

Traffic

The primary function of traffic nets is the transmission of formal written communications. Amateur radio operators have traditionally been responsible for relaying both routine and emergency messages on behalf of third parties as part of the public service purpose permitted by the regulations established by the governments of the United States and Canada controlling amateur radio. The primary goal of the American Radio Relay League, or ARRL, was to transmit communications from other parties. It was established in the early 20th century (1914–15). Except for North America, most nations and parts of the world forbid amateur radio operators from relaying communications on behalf of other parties.

Routine message handling has become much less popular and is largely used for education, as nearly everyone now has access to low-cost communication tools. During emergencies, including natural catastrophes, information is transmitted into and out of affected areas using traffic nets, which run on emergency-powered stations. This is especially important when traditional communication channels are either destroyed or made useless.

DX

The purpose of putting together DX nets is to assist amateur radio operators in making contact with stations located in remote areas or in regions that have a scarcity of amateur radio operators. A ham radio operator has a better chance of making contact with another station if they participate in a DX net than if they simply tune around the amateur radio bands in a haphazard manner and hope to hear something.

Club or Topic

"Nets" are often held by amateur radio clubs to promote contact among club members. These can be hobby groups or subject-specific clubs that were founded with a specific interest in mind. Internet forums known as "nets" are frequently established by non-club and special interest groups to facilitate talks about particular subjects. There are several different types of these nets in use today. One example of this is networks that get together to discuss vintage or antique radio equipment. Another example of this is networking to use and talk about AM voice transmission mode.

Finding and Participating in Ham Radio Nets

Amateur radio nets that have registered with the database of the ARRL Headquarters net directory are listed in the ARRL Online Net Directory. It mostly covers nets of interest to amateur radio operators in North America, particularly in the US and Canada. The database includes global coverage nets and maritime service nets in addition to National Traffic System Area and Region Nets. Marine service nets are also included. On their website, you can use the database search function to look for a specific net.

One of the primary focuses of the directory is on public-service-oriented nets that contribute to the ARRL National Traffic System (NTS) and the Amateur Radio Emergency Service (ARES) (ARES). You can acquire additional information regarding NTS and ARES activity in your region by contacting the leaders of your local ARRL Field Organization.

As "often scheduled on-air meetings of hams who share common interests," ham radio operators have traditionally participated in net participation. Presumably, the first net was put together as soon as the first two hams connected. Sometimes people gather over the nets for pure entertainment, discussing things like product collecting, radio chess, or earning rewards. Certain networks are designed with practicality in mind; these include traffic control, emergency response, and weather reporting networks. A net is said to be directed if it adheres to standard operating procedures. Almost all directed nets have an identical basic architecture. A net control station (NCS) is responsible for launching the net operations, preserving order, directing the activities that take place on the net, and subsequently ending the net operations in a methodical manner.

Stations interested in participating in the net listen for instructions from the NCS before checking in. A net manager collaborates with the NCS stations to ensure that the net meets regularly and sets the net's priorities and policies.

Checking in with Ham Radio Nets

You have to register both your location or status and call sign with the NCS before you may check in to a net. Verify that the Speaker is fully visible to you and that you can hear and comprehend what they are saying. You'll have to wait until the NCS requests visitors if you don't use the internet frequently. When you check in, please raise your hand once and pronounce your call sign phonetically if you are speaking to someone. If your call sign is not repeated the first time, you can ask the NCS to transmit it again or they can ask one of the listening stations to do so. Repeat if the NCS does not automatically replicate your call sign the first time.

There are several ways to stay in touch with the business (such as announcements) or traffic (messages) on the internet; the best way to figure out which is acceptable is to listen to the internet. The most common method is to say something like "N-AX with one item for the net." The NCS has acknowledged receipt of your submission; you are now awaiting further instructions.

What they say, "When in Rome, do as the Romans do," applies here. Pay attention to the checks that other network users are making. You can either wait until the check-in procedure is completed and the NCS calls for net business, or declare at the time of check-in that you are interested in contacting one of the other stations checking in as though it were a net business. If you opt for the latter option, you will need to wait until the NCS calls for net business. In addition, you can choose to check in using your call sign and reply, "One item." In either case, the NCS will ask the other station to acknowledge you before grouping the two of you together in accordance with the regulations for the net.

Activity

1. Describe the basic structure of the net.

2. What are the modes of net operation?

3. Mention two (2) types of amateur nets.

CHAPTER 11
PROGRAMMING WITH CHIRP

It is not sufficient to acquire a Ham radio and carry it in your backpack. Knowing who you want to contact and how to do so is an essential component of actually having communications prepared for an emergency. If you need to contact your local emergency services, for instance, you do not want to have to fight to remember the frequency 150.965 and manually enter it into your phone. It's better to have that number programmed into your phone.

The majority of radios, including those used for preparation like the BF-F8 and BaoFeng UV5R, come with an integrated digital address book. But they are about as user-friendly as a nineties VCR; manually adjusting the radio's settings via the keypad takes a lot of time and is highly unpleasant. By connecting your radio to your computer, you can use the free Ham radio programming tool CHIRP to manage the address book on your device. It works with the three most widely used desktop operating systems and is free to use: Linux, macOS, and Windows are all available; however, Windows is the better option if you have the option because it removes a few minor annoyances. For the most part, CHIRP works with radios. The entire list is available on their website. Although CHIRP's availability is fantastic—especially given that it is free and open-source software—using it might be annoying for those who wish to include ham in their preparations but do not wish to "be a Ham." All you'll need is a little perseverance and an FTDI cable that is compatible with your radio.

The following types of data formats can be processed by CHIRP:

- Comma Separated Values (.csv)
- Comma Separated Values generated by RT Systems (.csv)
- EVE for Yaesu VX-5 (.eve)
- Kenwood HMK format (.hmk)
- Kenwood commercial ITM format (.itm)
- Icom Data Files (.icf)
- ARRL TravelPlus (.tpe)
- VX5 Commander Files (.vx5)
- VX7 Commander Files (.vx7)
- Most popular modern amateur radios are supported by CHIRP via their interface cables.

Preparing to Program Your Baofeng Radio

You may get Chirp for Windows, MacOS, and Linux, so you are always covered regardless of the operating system being utilized. All you need to do is download the version that is appropriate for your system. Before you may program your Baofeng radio with CHIRP, you must utilize the programming cable after the software has been installed. Using the Baofeng USB programming cord is the best option. Make sure you buy a cable from a reliable vendor to avoid running the danger of certain procedures being left out. It is highly recommended you use a Baofeng-produced cable; those kinds of cables work perfectly with the included FTDI chipset. This cable is plugin play and ought not to require different drivers, but there is another option for you to also download drivers from websites that are compatible with Baofeng. Ensure you download and install the proper drivers for your USB cable.

Once the above settings and needs have been met, follow the steps below to program your Baofeng radio;

- Open/Run CHIRP, connect your Baofeng device and insert it into a USB port on your computer. You will first need to download a copy of the current configuration from your radio to your computer. Additionally, you will need to give CHIRP some details about your radio and the COM port (or, in Linux and Mac OS, which /dev/XXXX) that it is connected to. The channels that you have previously been tuned to should be listed after this step is finished.

Side Tip:

If you are utilizing a UV-5R variant, keep in mind that the configuration setting described here applies to a number of those variants (such as UV-5RE Plus). On the CHIRP website, you can find out which models are protected and which one you should select if you have any questions regarding this matter.

- Navigate to CHIRP, go to **File > Open Stock Config**, and choose the group of channels that you would like to add to your Baofeng. You can choose to make a choice of channels or choose all of the channels with the use of the hot-key combinations for your OS.

Windows (and most Linux systems unless you've remapped stuff)

- Select all = **CTRL + A** (keys on your keyboard)
- Select individual Channels = **CTRL + mouse click**
- Select Range of Channels = **Mouse click** to highlight start then, **SHIFT + mouse click** on the last in the range that you'd like to select

Then **CTRL + C** (or **EDIT > COPY**) to copy the selected channels

Side Tip

Importing data from other sources besides the stock configuration can be accomplished by selecting Radio from the radio drop-down menu and then selecting Import from other data sources.

RadioReference.com, RepeaterBook, and RFFinder are just some of the alternatives that may be available to you.

- Go back **to your Baofeng tab**. Select the first channel that has not been occupied then click on **Edit > Paste (or make use of the CTRL + V)** to paste in the previously copied channels from the Stock Config into your Radio's Config
    - If per adventure you notice you do not have enough channel slots, you can always add more channels. It is worth noting that Baofeng supports up to 128 memory channels; 0-127).

Side Tip:

If you want to import FRS or GMRS channels for receiving, you can disable transmission capabilities, but transmitting with a Baofeng while using FRS or GMRS is illegal. You can still import those channels to use for receiving, though. To proceed, choose "none" for the frequency at which you wish to stop TX broadcasts in the "Duplex" column. Next, select the desired option from the "off" drop-down menu. This will allow you to receive signals from those channels while preventing accidental transmissions on those channels.

- All you have to do in this step is to upload this config back to your radio. Click on **Radio > Upload to Radio** and then CHIRP will write it to your Baofeng's memory. The same configuration options that were employed for the initial cloning might be selected. Instead of a transmit indicator light, you will see a receive light. If you want to make sure the radio is set up properly. You should be able to navigate through and observe the additional frequencies if you switch to the Channel/Memory mode and move up to the A frequency.

Alternative options

- Follow the first step above to download your radio config. Download the config baofeng-freqs.csv. Ensure the file is saved in a place you can easily remember.
- Undergo the third step above to modify the number of channels and there will be a need for you to configure this to 60 or 61.
- In CHIRP, choose **File > Import** then import the file above (baofeng-freqs.csv) that you have saved somewhere.
- Undergo the fourth step above to write this config to your Baofeng.

Weather stations

If you would like to include weather frequencies; make use of the same process which is **Radio > Import from Stock Config > NOAA Weather Alert**.

It is worth noting that two configurations are quite important for weather channels;

- It is not advised that, unless you are a licensed meteorologist, you converse on weather frequencies. Therefore, be sure that each of your weather frequencies has the Duplex setting set to Off.

- To tell your radio to skip over certain stations during its scan, put an S in the "Skip" column. Your radio will stop at every weather station it can detect on the dial if you don't make any changes to it and set it to scan mode. Furthermore, your scan will slow at weather stations because they are constantly sending.

Local repeaters

CHIRP is quite helpful when it comes to entering Ham radio repeaters. A particular type of Ham radio station that searches for signals and then "repeats" them across a larger area is referred to as a "repeater". Usually, this station is perched atop a steep hill or mountain. Many repeaters require special setups, such as a PL tone, and a frequency offset (because they usually listen on one frequency and transmit on another). Programming is made incredibly easy with CHIRP since it not only allows for the manual entering of frequencies, offsets, and PL tones but can also import frequencies from several well-known online libraries.

- Choose **Radio > Import from Data Source > RepeaterBook > RepeaterBook Proximity Query** to connect to the repeaters that are located close to you.

This will open a dialog box where you will need to enter the search distance in miles, the available bands, and your ZIP Code. To boost the number of repeaters that CHIRP finds, keep the Band setting at All. Using CHIRP, you cannot import frequencies that are incompatible with your radio.

Emergency and first responders' frequencies

There are two types in this section;

- People can use these standard frequencies to call for assistance. Imagine that you are calling the emergency number 911; no matter where you are in the United States; you can call 911 and receive assistance. In a similar vein, the vast majority of boaters are aware that Channel 16 on marine radios is the standard call-for-help channel.

- The particular frequencies that are utilized by the first responders in your area, such as the police and the fire department.

Below are some of the emergency frequencies;

- The 2-meter calling frequency in the Americas is 146.520 MHz, 145.5 MHz in Europe, and 145 MHz in the Philippines, Indonesia, and Thailand.

- In the Americas and Asia, the 70-centimeter calling frequency is 446, while in Europe it is 433.5.

- Marine VHF channel 16 (156.8 MHz) can be imported into CHIRP via Radio > Import from Stock Config.

- Channels 1, 3, and 20 of FRS. Scroll down to learn how to use CHIRP to program FRS and GRMS channels.

Many attempts were made by preppers to create their universal channels. There are moments when setting up your local first responder frequencies might become very difficult. By manually scanning until you find what is active in your area, you can accomplish this. Additionally, you may manually search websites like RadioReference and transfer the data across. You can also decide to give a local radio station or even the police a call.

FRMS AND GRMS

Both the Family Radio Service (FRS) and the General Mobile Radio Service (GMRS) are examples of non-ham bands that have unique purposes. The FRS standard is primarily designed for use with walkie-talkies. A license is necessary to operate GMRS, which is essentially an upgraded version of FRS.

To put it briefly, only specially designed radios are permitted to use the FRS and GMRS radio frequencies. Using a regular Ham radio to transmit on specific frequencies is prohibited by law. However, the FCC's regulations permit you to use any frequency you choose for communication in a life-or-death situation. This is due to the possibility that you will want to be able to access as many frequencies as feasible. Even though there might not be any Ham operators in the vicinity, you might be able to speak with someone using an FRS walkie-talkie or GRMS radio.

- Choose **Radio > Import from Stock Configuration > US FRS and GMRS Channels** from the drop-down menu.

Just as you were given a list of frequencies for the other alternatives, you will receive one here. By this time, they should all be working with your radio. As is normal, you will need to make some adjustments to the channel numbers.

Activity

1. Highlight the steps to programming the Baofeng radio with the use of CHIRP

CHAPTER 12
HAND PROGRAMMING THE BAOFENG

Talk on the radio occurs on a frequency like 146.52. Since you will probably use several frequencies for radio communications, you can program these frequencies into your radio so that they become channels rather than learning a ton of new ones. This implies that all you need to do is remember to tune into channel 1 instead of the frequency if you program 146.52 into it. Instead of having to key each channel each time, you may also opt to browse through your list of channels.

When having to communicate just one frequency, this is known as simplex. For instance, This is just simple if you tune into 146.52 and there is someone else on the same frequency that you are speaking to. One of the most popular dual-band portable two-way radios available anywhere in the globe is the Baofeng UV-5R. It is inexpensive, offers a ton of functionality, and has a small footprint. For many users, especially those who are unfamiliar with programming portables, it is also one of the hardest to program. This is primarily because of a badly written user manual and nearly nonexistent support from Baofeng, the business that manufactures the product.

Programming refers to the process of adding different frequencies to different channels. When you employ repeaters, which are automated stations that listen for transmissions on one frequency and "repeat" them over another frequency, the significance of this factor increases. Repeaters can be found in several different forms. Repeaters introduce more complexities into the mix.

Many radios require that the radio send a distinct tone, sometimes known as a CTCSS tone or a DCS tone before the repeater will rebroadcast a signal from that radio. If you do not first program the correct tone into the channel, you will not be able to effectively transmit to that repeater. Repeaters use duplex communication because it uses two separate frequencies for transmission as opposed to simplex communication.

Programming Simplex Channels into Baofeng

Manually programming simplex frequencies is a sensible place to start because it is less complicated than manually programming repeater frequencies. You must first switch from channel mode to frequency mode to proceed. While you navigate between pre-programmed channels in the channel mode, you dial in frequencies directly in the frequency mode. Turn on your BaoFeng and have a look at the screen to get started. The channel numbers will show up on the right side of the screen when you are in channel mode.

- Pressing the **VFO/MR button** will toggle between the different modes.

You've probably noticed that there are two different sets of frequencies displayed on the screen, one on top and one below.

- By pushing the **A/B button**, the BaoFeng gives you the ability to swiftly switch between two different frequencies. To program a frequency, you will first need to make sure that you are on the top frequency, which is denoted by a small arrow on the left.

Use the keypad to enter the desired frequency, such as 146.52 MHz, once you have entered the frequency mode and are on the top frequency. The frequency you wish to input needs to have zeroes added to the end since the BaoFeng requires three digits to be entered after the decimal point, as demonstrated here: 146.520. As of right now, the decimal point is not needed, so just enter 146520 without it.

Press the **Menu button once** you have determined that the top frequency is the one you want to program. You can type 27 instead of scrolling to the MEM-CH menu item, which is located farther down the page. The MEM-CH setting is the one that is responsible for programming channels into memory. When you are on MEM-CH, pressing Menu will allow you to change that setting. When you have completed this step, you will see the small arrow on the left change from pointing to MEM-CH to pointing to CH-000, which is the default channel.

Note: You can choose to tap the Exit button at any time to leave the settings menu, either before or after the changes made have been saved.

You have the option of using the arrow keys to browse the channel or inserting it on the keyboard if you want to. Once you've selected the channel you want to program the frequency into, touch **Menu and then Exit.**

To navigate to channel mode, select **VFO/MFR** when you return to the main menu. To verify that the channel has been correctly programmed, either use the arrows to travel to the programmed channel or enter it into the keypad.

Programming Repeaters into a Baofeng

Once you have a solid understanding of channel programming, you are ready to install repeaters by hand. Repeaters need to know the following four pieces of information: the primary frequency, offset, offset direction, and tone. (The tones are shown as tone in / tone out in RepeaterList; however, for FM stations that are BaoFeng compatible, they ought to be the same.)

The primary frequency is the one on which the repeater transmits; the frequency you are listening to is the secondary frequency. The offset is the frequency to which you should communicate, and it is the frequency on which the repeater listens. An offset can be expressed as a positive or negative number. Thus, the frequency at which you should transmit is 146.07 MHz if the repeater's frequency is 146.67 MHz and the offset is -0.6 MHz. If the offset was +0.6 instead, you would switch to transmitting on frequency 147.27.

For instance, make use of W4CAT, which is a repeater around Nashville:

- **Frequency:** 146.955
- **Offset:** -0.6 MHz
- **CTCSS:** 114.8

Take the first couple of steps above again; ensure that you are on the top frequency in VFO mode and then insert the repeater frequency.

Configure the offset direction;

- Press Menu
- Type 25 or scroll to SFT-D
- Press Menu
- Use the arrows to set +, -, or off
- Press Menu

Configure the offset frequency:

- If you're in the menu, press 26 or scroll to OFFSET
- Press Menu
- Enter the offset (for 0.6, type 000600)
- Press Menu

Finally, the tone:

- If you're in the menu, press 13 or scroll to T-CTCS (short for transmit CTCSS)
- Press Menu
- Use the keypad to enter the tone frequency
- Press Menu

Once you are sure everything has been well configured, save the frequency to a channel just as you would for Simplex. The offset direction, offset frequency, and CTCSS tone settings ought to be saved to that channel.

Verifying whether or not everything was correctly saved is not difficult. Point the pointer to the channel that is being used by the repeater. An offset should be shown by two plus and minus signs at the very top of the screen. Two things ought to occur when you press the PTT button on the radio's side to start a transmission: The frequency should decrease or increase to the offset, and CT, which indicates that a CTCSS tone is being transmitted, should display on the left. For instance, if an offset of -0.6 MHz is applied to 146.955, the frequency moves to 146.355.

Deleting a Channel

It's much easier to delete a channel than add one:

- Press Menu
- Enter 28 or scroll to DEL-CH
- Press Menu
- Scroll to or enter the channel to delete
- Press Menu

Be careful here because there is no confirmation prompt. If you get into the menu and change your mind, press Exit before step 5.

Theory of Operation

There is space for a send frequency and its associated features in each memory channel, in addition to a receive frequency and its associated parameters. The other channel is eliminated along with the first. The receiving space is where data is saved when a frequency and its corresponding parameters are placed into an otherwise empty channel. The data you just input is saved in the transmit space when you restore a frequency and its related parameters to the same memory channel a second time. As a result, programming a duplex channel requires two separate processes because its transmitted and receive frequencies are distinct.

Putting away a simplex channel requires only the first step to be taken. Once the frequency and settings have been saved, they will continue to be maintained in the receiving space. You will have simple communication if you broadcast on a channel that has only the receiver space programmed in the radio. This will cause the radio to transmit on the receive frequency.

Stated differently, a channel is said to be in the simplex mode if its broadcast space remains empty and only its receive space is configured. It will send and receive signals on the receiving frequency. If the broadcast and receive portions of a memory channel are both programmed, the radio can operate in a duplex on the two separate frequencies.

Activity

1. How can simplex channels be programmed into Baofeng?
2. How can repeaters be programmed into Baofeng?
3. Briefly highlight the steps to deleting a channel.

CHAPTER 13
LINGO & TALKING ON THE RADIO

Have you been conversing via two-way radio and come across some radio jargon that you weren't quite sure what to make of? When you completed stating a crucial point, for instance, the person you were conversing with might have responded, "10-4," "Roger that," or they might have even asked you a question in a language you weren't acquainted with, like "Sure, what's your 20?" These are some fundamental instances of long-standing short-hand radio jargon that was developed to facilitate concise and understandable conversations for radio users. Unfortunately, because these terms are foreign to you, you wouldn't fully understand them. You'll learn some of the most common jargon used on the radio these days from this chapter.

Some radio expressions are probably already well-known to both radio users and non-users because of their widespread use in popular culture. These jargons include CB radio lingo that appears in songs and movies as well as police radio codes that are utilized on television.

Even if you are completely unfamiliar with radio communications, the following are some of the words that you will probably be familiar with.

- **Roger That**: A short and sweet approach to convey that you have heard and comprehended what the other person is saying. The name "Roger" comes from the days when people communicated using Morse code, and the letter "R" was used to signal that a message had been received or understood. The United States military took up the use of the term "roger" for the same reason as radio communications became more widespread and advanced technology was developed.

- **May Day**: A line that, with any luck, you will never hear in real life but will surely hear in the movies. It is a signal that is widely acknowledged as a cause for alarm and is generally considered to indicate a "life-threatening emergency." Generally speaking, the word "mayday" is used to indicate that a vehicle or form of transportation—such as an airplane, boat, helicopter, etc.—is about to crash. The origin of the expression "come help me" dates back to the early 1920s and is derived from the French term "m'aidez," which means "come to help me."

- **Over**: this word is often used at the end of a sentence or phrase to show that the person has finished speaking.

- **Out**: this word shows that the person is signing off.

- **Read/Copy**: these two words are used to inquire if the speaker is being heard or understood, for example, Do you read me? or Do you copy? Think of this as "Can you hear me"?

- **Wilco**: The literal meaning of this phrase is "will comply," and its use suggests that the speaker intends to carry out the activity that was requested of them.

Q codes used by Ham Radio Operators

Three combinations of letters that begin with the letter "Q" are known as Q-codes or Q-signals. To improve the speed and efficiency of radio conversations, hams employ these codes in opposition to ordinary language.

There are numerous Q codes, and each one has a distinct meaning. When a Q-code is accompanied by the question mark symbol, it can be interpreted as a question; otherwise, it can be interpreted as a statement.

Below is a list of the most common Q-codes you would have to know to learn ham radio lingo basically and ensure communication becomes as simple as possible.

QRL

Among the Q signals that are most commonly used is QRL. You have to send a QRL to find out f the chosen frequency is available for use before you can make a call.

'Are you busy?" is represented by the message QRL, which would be sent by the amateur radio station wishing to use the frequency. The station in charge of keeping an eye on the frequency should send out the QRL signal, which signals "I'm busy; please don't interfere," if t is already in use.

QSL

In the beginning, a QSL was used to confirm that a formal message had been received. In today's world, it indicates that you have obtained a transmission, which may or may not contain a message that is formal or casual. It is possible to transmit a QSL in the form of a card, an email, or even an audio recording. For instance, the QSL might read, "I acknowledge receipt." Can you confirm that you have received this?

QRP

If anyone says, RIG HR IS QRP, what the person is trying to say is that the power output of the transmitter is 5 W or less. Your transmitter output ought to be 5W or less for you to qualify for the QRP category. A candid example is; QRP (I will decrease power) or QRP? (Shall I reduce the transmitter power?)

QRU

Often used in radiogram communications, this Q-signal is sent and received by amateurs. Either they would ask the other operator if they had any messages to transmit via this signal, or they would let them know that they don't have any to transmit. These days, you might hear radio stations state, "I AM QRU," which means they plan to wrap up the conversation and would like to speak with you again only if you have anything more to say. They most likely also indicate this if they transmit QRU. Saying QRU: I don't have anything for you is one way to express it, alternatively, you may say QRU. Have you got anything to give me?

QRS

This code can be used to indicate to the other ham that you would like to broadcast slower codes or to ask if they would like to send slower codes. It would just be polite if you could take your time sending this code.

QRV

Initially, this simply means that a station is ready to get a message copied. Nowadays, however, shows that a station is ready to go on air or make inquiries if one is prepared to go on air. For instance, QRV?: " Are you ready?" QRV: "I'm ready.

QRT

This code alerts the other frequency that you are going to go off the air and asks if it is all right to get off the air. When used as a statement, it conveys the meaning of "cease sending." As a follow-up query, "Should I cease sending?"

QSO

This code is often used in the form of a noun to mean a contact. For instance, "I can communicate with QSO directly" or "Can you communicate with QSO directly?"

QRZ

A radio operator can use a Q code known as QRZ to enquire, "Who's calling me?" or indicate that a certain individual is calling. For example, if someone says "QRZ [name]," it implies "You're being called by [name]."

QTH

"My location is _____" is what "QTH" stands for. Hence, if you give the other operator the message "QTH Florida," you are informing them that your location is in the state of Florida. When a question mark is placed after the Q-signal, it becomes a question and is written as QTH? The inquiry that is being asked is, "Where are you located?"

Communication Terms

Anytime you check into a repeater or amateur radio band, or you are listening to the ham radio band with the use of a scanner, you will find lots of lingoes you might not understand. Here are some of the commonly used lingos and what they mean;

73

This phrase translates to "best wishes" or "best regards." It's a very polite way to terminate a conversation. When concluding or wishing to say farewell, you can use it.

88

"Love and kisses" is what the term is used to say. It is somewhat more private than 73 and frequently utilized by family members and close acquaintances. Although using it with strangers is not inherently improper, you should exercise extreme caution if you do so or if you do not have a friendly relationship with them.

DX

"D" stands for "distance," and "X" denotes an unknown value. It is a reference to a station that is hard to hear and is located far away on a frequency. The term can be used either as an adjective, like in "that was a DX QSO," or as a verb, like in "I'm DXing today." DXers are enthusiasts of amateur radio who enjoy listening to stations that are broadcasting from a considerable distance.

99

The Q-code is used to signal that the discussion is about to conclude. It is customary for one person to say "99" and the other to respond with "73." You may say something like, "I have to go now, 99," for example. After that, the other side would reply with the number "73."

XYL

A wife is referred to as an XYL, which is an abbreviation that stands for "ex-young lady." You may, for instance, state anything along the lines of "I wanted to buy that new radio, but the XYL claimed it was too pricey."

YL

A young lady who is a licensed amateur radio operator is referred to as a "YL," which is an acronym. For example, you could say something like, "The YL on 15 MHz was very nice and articulate."

COPY

The word "Copy" comes from the word "Understood." It demonstrates that the recipient has grasped the meaning of the message that was sent to them. You can make use of it if you overhear or receive communications between two stations that are beneficial to your station.

Phonetic Alphabet

Have you ever been unable to figure out which letter someone was saying? All of the various letters have a similar sound when uttered. It is quite easy to mix up the letters "M" and "N" or "B" and "D," especially when using an electronic device to communicate. The phonetic alphabet, commonly referred to as the spelling alphabet and consisting of a string of phrases that identify each letter, is widely used by people who speak over the radio to refer to letters. This is being done in an attempt to solve the issue.

Police officers typically use the police phonetic alphabet to communicate a license plate number. Officers may convey the number 111-ABC, for instance, as "1-1-1-Alpha-Bravo-

Charlie." Throughout the world, the International Phonetic Alphabet—also referred to as the military phonetic alphabet—is used in many sectors, including public safety, education, healthcare, and even manufacturing. The International Phonetic Alphabet is another name for it.

Below is a complete list of the alphabet;

Alpha, Bravo, Charlie, Delta, Echo, Foxtrot, Golf, Hotel, India, Juliet, Kilo, Lima, Mike, November, Oscar, Papa, Quebec, Romeo, Sierra, Tango, Uniform, Victor, Whiskey, X-ray, Yankee, Zulu.

10 Codes

Ten codes provide a simplified radio communication system that may be used by a wide range of customers and industries. Your chances of hearing a 10-code are no greater when working in public safety than they are in the industrial industry.

To put it briefly, 10-signals, or 10-codes, are just numerical representations of phrases.

The following is a list of some of the most common 10 codes and what they signify:

- **0-1:** Bad reception
- **10-4:** "OK" or "Affirmative," similar to "Roger"
- **10-9:** "Say again", or "repeat, please"
- **10-20:** Location, as in "What's your 20?"
- **10-36:** Current time, "Can I get a 10-36?"
- **10-69:** "Message received," again, much like "Roger"
- **10-77:** Estimated time of arrival, "Alpha 10-77"

10-codes, like "Roger" and "Mayday," date back to the first half of the twentieth century. In the 1930s, Charles "Charlie" Hopper (District 10) was the communications director for the Illinois State Police and is credited with inventing the codes. Due to limitations in radio technology at the time, there was a brief delay between when an officer pressed the talk button and when their voice was transmitted. Hopper realized that putting a "10" before the codes gave the radios enough time to catch up, ensuring that both complete and abbreviated messages got through.

Additionally, truckers have eleven codes of their own, some of which are specific to them and have connotations similar to those of law enforcement. The standard meaning of "10-4," for instance, is "I understand." Many contend that the 10 codes are no longer relevant because their meanings vary among departments, regions, and industries. Indeed, a lack of uniformity among first responders and law enforcement has had a terrible effect on communication and coordination in the wake of natural catastrophes like Hurricane Katrina.

Departments have been urged by officials, particularly those affiliated with the Federal Emergency Management Agency (FEMA), to substitute "Plain Talk" for technical terms while communicating over the radio. In federal communications, plain English is taking the place of codes. Plain Talk proponents contend that although it may take longer to communicate ideas, the extra effort is worthwhile to guarantee interoperability and mutual understanding.

The issue is still being debated, and the 10 codes are still widely used in public safety, with an official guide created by the Association of Public Safety Communications Officials (APCO).

Activity

1. Mention five (5) of the q codes used in ham radios.
2. What are phonetic alphabets and how can they be used with ham radios?
3. Highlight three (3) of the communication terms in ham radio.

CHAPTER 14
FINDING A CLUB

The many clubs within the amateur radio community are highly valued. Because of the lectures they conduct and the testing sessions they monitor, they are a major supplier of new hams. By joining a club, radio amateurs can get information from other members about new laws and regulations, new technology, and new amateur radio products. Another crucial avenue for radio amateurs to network with one another is through clubs. Finally, clubs offer a critical social outlet to members of the radio amateur community.

Joining a local ham radio club is the fastest and easiest way to connect with other ham radio operators. The history of ham radio is almost as old as that of ham radio groups. The initial clubs were nothing more than a group of enthusiasts who collaborated to build radios during a period when the technology was still developing and success was far from certain.

Ham radio clubs are great opportunities to meet more seasoned people and get advice and assistance. You'll find that as you get started with amateur radio, you have several basic questions that need to be answered. Membership in a general interest ham radio organization is always the first step that should be taken. Discovering one that prioritizes offering assistance to inexperienced hams would be really beneficial. You'll find that the road to enjoying ham radio is a lot less rocky when you travel with others.

The vast majority of amateur radio operators are members of at least one amateur radio group. Many individuals are part of one group with a broad interest and two clubs with more specific concerns. The vast majority of ham radio groups, whether they are local or regional, hold meetings in person. The majority of members come from just one geographical region.

Specialty ham radio clubs are primarily focused on the activities of their members. Activities such as amateur television, contesting, and low-power operations have the potential to attract a significantly bigger membership. It is possible that certain ham radio club chapters do not hold meetings in person. They might decide to have virtual meetings exclusively.

You may find local ham radio groups by just doing an online search for them. The amateur radio club American Radio Relay League (ARRL) has a database of related organizations on their website. To obtain a list of local ham radio clubs, just enter your zip code, state, or city. Focus on general interest clubs and look for clubs that welcome new hams as members.

Should there be multiple ham radio clubs in your area, find out when they meet to see which one has the most convenient meeting times for you to attend. See if you have any specialized hobbies or if any of the local clubs dedicate any of their time to your particular area of interest. Maybe the best course of action is to go to a few meetings of each club to see which one has the most interesting activities for you.

You will quickly realize that the challenge is not in locating clubs; rather, it is in selecting the ham radio club that is the most suitable for you. You are free to join as many ham radio clubs as you choose so long as none of them require a significant personal commitment on your side like public service clubs do. If this is the case, you should only join a single club. You can gain useful insight into one of Ham Radio's many subfields by reading the newsletters and browsing the websites that are maintained by the majority of the hobby groups.

Following your selection of a general interest ham radio club, you should show up to meetings, make introductions, and start participating in club events. Give your time to help set up before meetings and clean up afterward because you will get out of the ham radio club exactly what you put in. You may make friends for the rest of your life with the people you meet.

Reasons to Join a Radio Club

- **Elmers Everywhere**: Entering the world of amateur radio without first seeking out a mentor would be equivalent to plunging headfirst into the deep sea without a life preserver. For many new Hams, the knowledge and direction of an Elmerv0 can make all the difference between sinking to the bottom and floating along happily. Acquiring membership in an Amateur Radio club gives you access to many and probably a community of seasoned Elmers who can help you get started. Additionally, clubs frequently have in-person or online lectures on pertinent issues related to amateur radio. If you are a new Ham, signing up to be a member of a local club can help you get on the air more quickly and successfully than if you try to do it alone. The majority of clubs have members who are quite knowledgeable about setting up technology to function as you wish. The learning curve for configuring a functional station will be significantly lowered when you pick up tips and tricks from more seasoned hams. You'll also become adept in the essential operational abilities.

- **Serve the Community:** Many organizations provide their members the chance to take part in community service projects where they handle communications for neighborhood gatherings including concerts, bike rides, walks, runs, parades, and other events of a similar nature. By making use of your license to do so, you may also learn more about how to begin offering emergency communication services. Furthermore, you will have the chance to take part in activities that promote community awareness of the advantages of Ham Radio.

- **Meet Cool People**: Talk to any Ham and they will tell you that they have made some of the best friends of their lives through their hobby of amateur radio. Do you think that you are the only inexperienced operator in the group? Consider this again. The club's membership spans all skill levels, from those who have completed their first QSO to those who have been inducted into the DXCC Honor Roll.

    "The majority of active Hams are members of many clubs. Although Amateur Radio is inherently a "technical" hobby, the fundamental purpose of the technology is to enable people to communicate with one another. If you are a member of a local club, you will have the opportunity to get to know the other Hams in your area by communicating with other club members.

- **Stay Up To Date**: Amateur radio clubs are great sources of information on anything going on in the Ham radio community, both locally and globally, through meetings, bulletins, and websites. Members of a club provide you with up-to-date information on events, gear sales, competitions, and much more. These people are people you know and have a good rapport with.

- **Broaden Your Horizon**: Joining an Amateur Radio club can provide you with an introduction, or a reintroduction, to the myriad facets of the hobby, regardless of whether you are a novice or an experienced Ham wishing to get back on the air after a period of inactivity lasting many years. Individuals who discover a particular field of interest may later choose to become members of more specialized groups, such as AMSAT (Radio Amateur Satellite Company) or QRP ARCI (QRP Amateur Radio Club International).

"Getting involved in a local club will assist you in realizing the range of activities available to you as a Ham, including DXing, contesting, fox hunting, moonbounce, VHF/UHF, QRP, satellites, and other activities.

Joining a local club will also provide you the motivation to go up the license classes and become an Amateur Extra or General, giving you access to the most operating privileges possible across all bands.

- **Enjoy Yourself**: For most ham radio fans, the answer to the question "Why are you a Ham?" is rather simple. since it's fun! As a member of an amateur radio operator club, you will find that the hobby is much more enjoyable. Two occasions when club members can foster a sense of camaraderie and fellowship with one another are picnics and field days.

It was all it took to become involved with a group that was open to exploring and trying new things for them to have the time of their lives participating in all the activities and areas of interest offered by an active local club for amateur radio operators, or Hams, as they are known to have done. Getting involved in the activities of a local amateur radio club is the greatest way to fully commit to the hobby of amateur radio.

Note: Most clubs for amateur radio demand that all of their members hold valid amateur radio licenses. You should consider the many membership types that your club plans to provide in the future, along with the benefits that come along with those memberships. It is also up to you to decide if membership dues or fees will be required in order to become a member of your club (your by-laws can specify the precise amount of these dues or costs).

Activity

1. Briefly highlight five (5) reasons you should join a Radio club.

CHAPTER 15
BASICS ABOUT COAX & YOUR BAOFENG

The coaxial cable which can also be called Coax for short can be described as a type of electrical cable that has an inner conductor surrounded by a concentric conducting shield with both differentiated by a dielectric material; most coaxial cables also have a protective outer sheath outer jacket or sheath. The term points to the inner conductor and also the outer shield which share a geometric axis.

Coaxial cable is one type of transmission line that is used to convey high-frequency electrical signals with the least amount of signal loss. It is used in many different applications, such as high-speed computer data buses, cable television signal transmissions, telephone trunk lines, and radio transmitter and receiver antenna connections. The accurate and constent conductor spacing is achieved by precisely controlling the dimensions of both the cable and connections, setting it apart from other shielded cables. It cannot operate as an effective transmission line without this conductor spacing. This feature is absent from other insulated wires.

Radiofrequency signals are transmitted through a coaxial cable used as a transmission line. Applications for this technology include the use of feed lines to connect radio transmitters and receivers to their antennas, connections to computer networks (such as Ethernet), digital audio (S/PDIF), and the distribution of cable television signals.

In a perfect coaxial cable, the electromagnetic field that transmits the signal only occurs in the area between the inner and outer conductors. One benefit of coaxial over other kinds of radio transmission lines is this. This means that, unlike other transmission lines, coaxial cable runs can be installed near metal objects like gutters without resulting in power losses. A coaxial cable is also used to insulate the transmission from external electromagnetic interference.

Options for coaxial cables impact cost, size, attenuation strength, flexibility, and frequency performance. Although stranded conductors are more flexible, solid conductors are still required for the inner conductor. The inner conductor may need to be plated in silver if a higher frequency performance is required. It should be noted that the majority of the time, inner conductors for cables used in the cable TV sector are made of copper-plated steel wire. Braided copper wire is used in the majority of traditional coaxial cables to create the shield. Because the braid cannot be flat, this allows the cable to become more flexible, but it also means that there are holes in the shield layer and that the inner dimension of the shield fluctuates slightly. There are times when the braid is plated in a silver color.

Although it is currently more typical to have a thin foil shield coated in a wire braid, some cables have a double-layer shield for much-improved shield performance. The shield may consist of roughly two strands. Certain cables, such as the quad shield, which uses four separate layers of foil and braid, may invest in more than two shield layers. Certain shield designs sacrifice flexibility in favor of increased performance; these shields are composed of a single metal tube. These cables cannot be bent sharply because doing so will cause the shield to kink, which will result in signal loss. When a foil shield is utilized, the process of soldering the shield termination is simplified by the incorporation of a small wire conductor into the foil itself.

Types of Coaxial Cable

Hardline: Hardline is frequently utilized in radio communications, including broadcasting. This type of coaxial cable can be defined as one that is shielded by circular copper, silver, or gold tubing, or by a mix of these metals. Aluminum shielding is used by a specific hard line that is of poorer quality. On the other hand, unlike silver oxide, aluminum oxide often oxidizes easily and loses conductivity more quickly. As a result, every connection must be airtight and waterproof. Copper-plated aluminum or solid copper may be used as the center conductor. Copper plating provides sufficient surfaces to function as an effective conductor because the radiofrequency (RF) skin effect is a problem. Most varieties of hard lines utilized for external chassis or when exposed to the elements have a PVC jacket; however, certain internal applications might omit the insulation jacket. Hardline can be very thick; about a half inch or 13 mm and up to several times that, and also has low loss even at very high power.

Radiating or leaky cable: This is an additional type of coaxial cable that has a design akin to hard-line, but it has tuning slots split into the shield. These slots are set to operate in a particular radio frequency band or at a particular RF wavelength. When providing a tuned bi-directional preferred leakage effect between the transmitter and receiver, this kind of cable is frequently utilized. It is frequently used in subterranean transit tunnels, elevator shafts, US Navy ships, and other locations where using an antenna isn't practical.

Triaxial cable: this is a coaxial cable that has a third layer of shielding, sheathing, and insulation. The outer shield, which is oftentimes earthed, protects the inner shield from electromagnetic interference from external sources.

Twin-axial cable: this is a balanced, twisted pair around a cylindrical shield. It enables nearly perfect different signaling which is shielded and also balanced to pass through. Note that multi-conductor coaxial cable is also used sometimes.

Semi-rigid cable: The design is coaxial, and the outside sheath is composed of solid copper. Especially at higher frequencies, this coax offers superior screening than braided cables. The primary disadvantage is that, contrary to what the name would suggest, the wire is not designed to be bent after initial creation and is not very flexible.

Use conformable cable in place of semi-rigid coaxial cable when more pliability is required. The conformable cable may be made and stripped by hand without the need for any other tools, just like standard coaxial wire.

Rigid line: A coaxial cable with two copper tubes maintained in precise alignment every two meters by PTFE-supported couplings is called a rigid line. Because stiff lines cannot be bent, elbows are necessary. Two lengths of rigid lines are joined together using a flange or connection kit and an inner bullet/inner support. Stiff lines are connected using standardized EIA RF Connectors, whose flange and bullet sizes match the standard line widths. For every exterior diameter, inner tubes with a resistance of either 75 or 50 ohms can be provided. Connections are made using a more durable, rigid line with weatherproof flanges on outdoor antenna masts, etc. Indoors, rigid lines are frequently used to connect high-power transmitters and other RF components. To conserve weight and money, aluminum is commonly used for the outer line on masts and other structures; nevertheless, special care must be taken to prevent corrosion. A flange connection can also be used to join two hard lines together. The flanged rigid line interface is used by many broadcast antennas and antenna splitters even when connected to hard lines and flexible coaxial cables. A stiff line in many diameters is produced.

Coaxial Cable and Baofeng

Coaxial cables are very popular choices due to their shielded designs which give room for the center of the conductor to transmit data quickly while also being protected from damage as well as interference.

Coaxial cables are designed on the four layers below;

- A central conductor, typically a copper wire, through which data and video travel.
- A dielectric plastic insulator encircles the copper wire.
- The cable is then shielded from electromagnetic interference by a copper braided mesh (EMI)
- The external layer is a plastic coating that guards against damage to the internal layers.

A coaxial cable's core conductor transmits the data, and the surrounding layers of shielding help to reduce electromagnetic interference (EMI) and prevent any signal loss, also known as attenuation loss. In addition to offering some insulation, the first layer, usually referred to as the dielectric, is in charge of establishing some gap between the core conductor and the outer layers.

Types of coaxial cables that can be used with Baofeng radios;

RG-6 Cables

RG-6 cables deliver signals of higher quality even if their conductors are larger. They can withstand GHz-level signals more successfully because they are made with a different type of shielding and have thicker dielectric insulation. This specific wire is extremely thin, making it easy to install in walls and ceilings.

RG-59 Cables

The RG-59 cable and the RG-6 cable are fairly similar; however, the RG-59 cable's center conductor is much thinner. In-home settings, this statement is commonly utilized. It is a great choice for transmissions at low frequencies and across short distances as a result.

RG-11 Cables

Due to its bulkier design, the RG-11 cable can be easily distinguished from other types of coaxial cable; nevertheless, this also makes handling it more difficult. However, compared to

RG-6 or RG-59, it exhibits less attenuation, suggesting that it could be able to transmit data over longer distances.

Activity

1. What is a Coaxial cable?
2. Mention three (3) types of Coaxial cables.
3. Differentiate between the Coaxial cables used in Baofeng.

CHAPTER 16

ARE YOU GOING TO USE THAT RADIO?

Using a Baofeng VHF radio provides several perks, including the option to take a split channel. This can be a good approach to staying organized and keeping track of the various talks that you have been having. While employing the use of a split channel, you can carry on separate conversations on either side of the radio. This can be useful in situations in which you need to be able to concentrate on one conversation while simultaneously being able to hear another.

The frequency ranges that the Baofeng is sensitive to are 136–174 GHz and 400–520 MHz. It is, however, restricted to FM reception only. It is commonly known that the Baofengs can operate on any frequency between 136 and 194 MHz and 400 and 520 MHz. It is legal for you to own and operate such equipment as an amateur radio operator. The BF-F8HP (High Power) is the strongest BaoFeng radio currently on the market. A reputable manufacturer produces a high-quality ham radio, the Baofeng UV-5R 65-108. The frequency bands of 700 and 800 MHz are frequently used by police radios. FRS frequencies can be used on GMRS radios in addition to GMRS frequencies without a license. If you're an amateur radio operator looking for a commercial walkie-talkie, go no further than the Baofeng UV-5R, which features dual-band radio and dual-mode SAR capability.

To listen to civilian aviation frequencies, look for a radio with an air band between 118 and 136 hertz. An airband radio's power output can range from 5 to 8 watts, and its average range is 200 miles. The frequency 121.5 MHz is still available for use in aerial scenarios. SAR is a voice communication technology used by the US Armed Forces in their 138.78 GHz military radio systems. Directional discovery is another application for this frequency (DF). A BaoFeng radio can transmit signals up to 15 kilometers under perfect circumstances. The MXT400 MicroMobile GMRS Radio is the only option to consider if you're searching for a long-lasting fix.

Baofeng two-way radios have continued to show the main and most reliable form of communication for thousands of workers and lots of people around the world. From industrial workers to police officers, Baofeng two-way radios characterize and embody a high level of reliability and durability to workers today. Below are the advantages of making use of the Baofeng radio;

Cost of Ownership

The Baofeng radios are extremely cheap and do not have a monthly maintenance or service fee. Without a need for the payment of fees, you significantly lower your monthly operating costs by a very huge margin. As against the use of cell phones that attract a monthly fee, the Baofeng radios offer you a great deal for your money.

Enhanced worker safety

Two-way radio use is highly beneficial, particularly when worker safety is considered to be of utmost importance. Emergencies are manageable promptly. Many individuals have stated that Baofeng radios' safety features allow them to shield workers from harm.

Battery Life and Management

A considerable number of mobile companies have manufactured phones with incredibly short battery lives, meaning that a phone cannot even last an entire work shift. One of the main requirements for significant commercial communication equipment is powerful battery performance. Baofeng radios have a longer expected lifespan of proper operation than a day. You will have enough power to last the entire shift with this amount.

Durability and Reliability

Baofeng radios are distinguished by their design versatility and portability. This lengthens the item's lifespan by lowering the possibility that it will be damaged. You end up saving money on repair costs when you use Baofeng radios. The results of a study by VDC indicate that just 4–8% of two-way radios are expected to go out of service in the next year, compared to 18–20% of cell phones. A malfunctioning mobile phone could cause job disruptions, which would include monetary losses.

Audio Quality

An abundance of noise is present in industrial and manufacturing settings. Speaking with each other is much more difficult when there is a lot of noise around. That being said, it is imperative that employees can hear each other well in this environment. In an environment where there is a lot of background noise, cell phones become useless because people cannot communicate with one another. To ensure that users can understand and be understood by one another, Baofeng two-way digital radios utilize tried-and-true software algorithms that lower background noise and increase speech transmission quality.

Value-adding Features and Functionality

Contemporary two-way radio Digital radios, such as the Baofeng two-way radio, offer new high-end integrated functions such as messaging, FM radio, and GPS. Other integrated features include GPS. Other recently developed cutting-edge applications for the workplace, such as location tracking, are an additional perk that the radio offers in addition to these advantages.

The Federal Communications Commission (FCC) has approved the Baofeng UV-5R radio in compliance with part 90 of the FCC's rules. The FCC only certifies dual-band amateur radios built by Icom, Yaesu, and Kenwood under part 95, and radio manufacturers block these devices to prevent them from using commercial frequencies for transmission. This is due to the FCC's prohibition on dual-band radios using frequencies intended for commercial use. The National Weather Service or your local volunteer fire department's dispatcher can be set to only receive messages. Owning a radio from any of the major brands comes in quite handy during emergencies.

The Baofeng UV-5R radios can be configured to accept amateur radio frequencies in compliance with the amateur FCC part 95 standards, or they can be used in compliance with the commercial FCC part 90 laws. Obtaining the required license from the FCC is essential if you want to operate your business in compliance with FCC part 90. To use the Baofeng radio for amateur radio operations, you will require a radio license from the FCC, at the very least for the Technician level.

One of the best things about Baofeng radios is that they are the most affordable models on the market. These radios are available online at sites like Amazon.com and Wish.com, or you may get them from a Ham radio dealer in your area. You may get these radios for as low as $50, or maybe even less. An external microphone, an additional rechargeable battery, or a case for six AA batteries are available extras. By taking off the Baofeng radio's antenna and swapping it out with a magnetic one, you can turn it into a low-power base station or a low-power car mobile unit. Any setup will function.

Activity

1. Mention four (4) benefits of using the Baofeng radio.

CHAPTER 17
NON-HAM COMMUNICATIONS

For non-commercial, personal communications that do not rely on a network or grid, amateur radio is not just at the top of the pyramid; it makes up around 95% of the pyramid. Your alternatives will be very limited, but they won't be nonexistent if you don't own a ham radio. And what about the five percent of the population that remains? It's possible that you don't consider yourself to be a ham and never will be. Or maybe you are a ham who wants to learn new skills. What else can be discovered here? What options are there?

The encouraging news is that non-ham communications offer a variety of options to choose from. Every one of them is low-cost and quite simple to put into action. Although, you will not be able to speak with someone located far away using any of these methods.

Multi-Use Radio Service (MURS)

Around five distinct FM channels in the VHF band at 151 mHz are used by MURS. It doesn't need a license for you to operate it. The maximum transmit power is two watts. The antenna's maximum elevation is 60 feet above the ground, or 20 feet above the structure it is mounted on. MURS can also be used for non-voice communications, like security systems and motion sensors. There is a good chance that there will be a rivalry for the restricted band space with just five channels being used. One other thing to note is that MURS was in the VHF business band before. The grandfathering of commercial business licenses issued to MURS frequencies allows them to continue using the band even though their equipment may far surpass MURS technical requirements. Business customers who are grandfathered in have some advantages over unauthorized MURS stations. As opposed to radios made for other services, those made especially for MURS usually have a higher price tag. MURS frequencies are included in a wide variety of radios intended for amateur license holders. This is allowed as long as you make sure you follow the wattage limitations on transmitters. Most amateur radio equipment automatically surpasses two watts unless the power is actively decreased.

General Mobile Radio Service

Between 462 and 467 MHz, GMRS uses roughly thirty FM channels for operations. If you are in the USA, you will need to get a license, which costs roughly $70 and has a 10-year validity period. GMRS and Family Radio Service (FRS) share 22 channels. A maximum transmitter power of around 50 watts can be achieved using GMRS, except on channels 8–14, where the maximum is one and a half (0.5) watts. Provided that the input and output frequencies match defined divides, operators may opt to use repeaters with GMRS. Simplex communications can use the frequencies in between channels 1 and 7, although they are often restricted to five watts. On interstitial frequencies between channels 8-14, simplex is also allowed although the transmit power limit is still 0.5 watts.

True GMRS equipment can be highly expensive. It's important to note that manufacturers often promote FRS radios as "GMRS radios." This is theoretically correct because the two services share frequencies, but you should study the fine print to make sure you know exactly what you are getting. Because FRS radios are so inexpensive, they are infamous for having poor construction quality.

Family Radio Service

The frequency range of the FRS corresponds to 22 of the GMRS's channels. FRS does not require a license of any kind which is a little difficult to understand. Why is a license necessary for GMRS but not for FRS since the two services share the same frequencies and are practically the same thing anyway?

There does not seem to be an explanation for the reasoning behind the actions of the government, but an examination of the more technical aspects of FRS might offer some insights. The Family Radio Service is limited to transmitting at a power level of two watts on stations 1-7 and 15-22. Channels 8 through 14 have a maximum power output of 0.5 watts. Why? It appears that there is no rationale. FRS had a 0.5-watt maximum power output cap across all 22 channels before the implementation of a 2-watt upgrade in 2017. FRS radios can only be used with original equipment factory antennas that are fixed to the radio permanently. It is against the law to alter or modify the antenna in any way. Moreover, the use of repeaters is not permitted. So, even though GMRS and FRS appear to be almost the same thing, the former has power and antenna limits that render it significantly less useful than the latter.

Citizens' Band Radio

CB radio was the social media of its day if you aren't old enough to remember when it was around. The channels were active with a conversation at all hours of the day and night. Even though it is a shadow of what it once was, the band is still performing there, and the instruments and other musical equipment are open to anyone who wants to use them. It is the non-ham communications method that has been around the longest and is the most well-known.

The forty AM channels on a CB radio are spread between 26–27 mHz. Getting a license is not necessary. Four watts is the maximum power output for AM, and twelve watts is the maximum power output for SSB. The use of antennas is unrestricted. Transmissions over CB radio are often subject to noise and interference, and the antennas needed for CB radio are usually larger than those needed for MURS or GMRS/FRS transmissions. CB is inexpensively accessible. You may get used radios for as little as $5.00 that are still functional and in decent shape. You can get CBs for very affordable prices, even if they are brand new.

Even though a large percentage of the activities on modern CB is illegal, the FCC has all but given up on enforcing the law and will only pursue the most egregious of violators. The FCC rarely takes legal action against a citizen for allegedly breaking the laws that apply to their band. Additionally, CB is well recognized for serving as a meeting place for people whose ways of doing things are best described as extremely strange and annoying, but not necessarily unlawful. On the Citizens Band, it is also forbidden to use ionospheric "skip" to operate DX (the authorized maximum range is 160 miles). You might probably guess that the laws of physics do not adhere to the regulations that humans make, which is why skip happens regardless, and many CB operators make use of it.

From all of this, it can be concluded that anyone wishing to use CB for legal purposes will have to come to terms with the fact that they will be sharing the band with a lot of weirdos and pirates. All things considered; CB is a very valuable alternative to ham radio for communication purposes.

The question "How far can I talk?" is frequently posed by non-hams. Or, they express their dissatisfaction when the pair of toy-like portable FRS radios they purchased from the sporting goods department at Walmart for the price of twenty dollars won't actually send a signal as far as thirty-five miles, as the label on the package boasted loudly.

You will not be able to communicate reliably more than a few miles at most on any of the services mentioned in this chapter, except for GMRS radios with a power output of 50 watts. That falls into the higher range. It only gets worse after that. The CB radio may have the most potential for broadcast distance after highly powered GMRS. When using a CB radio, you should anticipate a range of about three to five miles; however, this might expand to ten or twelve miles with base stations or SSB.

FRS is the least effective of the group because of the restrictions placed on the antenna. Simplex communication with handheld radios has a maximum range of one mile or less, generally speaking, independent of the technology being used. While it's possible that under ideal circumstances you'll perform better, you shouldn't count on it. Naturally, bases and mobiles will have a greater range. You will need to invest in a robust GMRS system or potentially a MURS system if you require reliable communications over a distance greater than five miles. There's little doubt that using handheld radios alone won't get you very far.

Under no circumstances should you take the range claims claimed by radio equipment seriously. Oh my gosh, the "35-mile" brag that's commonly heard on FRS radios is so ludicrous that it should be illegal. The majority of individuals have dabbled with FRS radios, but they have never been able to utilize them for communication beyond a few hundred yards. "Who gives a damn?" is presumably what's going through your mind if you are already well-known for your hamminess. Now that I'm an amateur, I have a lot more options, so why should I need any of these? These substitutes are an excellent option to preserve friendships and interactions with loved ones who do not prefer ham. It is not reasonable to anticipate that everyone will act like a ham. If you are in charge of creating a neighborhood watch, a prepper/survivalist club, or in charge of hosting a public event, it can be quite helpful to have a few radios that are plug-and-play and do not require a license to hand out to the attendees.

Maintaining unlicensed services for others while doing your own business over the ham bands could provide an additional layer of security for your SHTF operations. Suppose you have a stock of CB radios in storage that are all tested and operational and can be utilized with simple dipole antennas. You can give them to your neighbors in the event of a disaster. Though all of your important messages will be on the ham bands, which they won't be listening to, you will be able to listen in on their chats and respond as needed. You can also decide to keep your distance from the people in your community while still participating in their social circles. Nobody will think of you as the guy with all those weird electronics and antennas, and no one will know what he does with all that gear. You will, instead, assume the part of the considerate neighbor who sets up a communications system for everyone. You will be able to divert attention from yourself if you can help other individuals without authorization.

Finally, remember that non-ham conversations are just more options available to you. You would miss out on the several megahertz of band space that unlicensed radio services together make accessible if you limited yourself to just using ham frequencies. Despite its shortcomings, it nonetheless plays a part in the communications plan you've created.

Activity

1. Describe two non-ham communication radios and how you can make use of them.

CHAPTER 18
GETTING LICENSED & LEGAL

In the United States, for the majority of its parts, all radio transmissions fall into one of the three categories;

- The operator is permitted to send on that frequency
- The radio is permitted to transmit on that frequency.
- The transmissions are quite very low in power.

The easiest way to be able to make use of the radio legally or in general to make use of radio transmitters for personal communication purposes is for you and probably members of your family to obtain amateur radio licenses and also obey the needed rules.

Getting Your Ham Radio License

Basic Requirements
- Must possess a valid Taxpayer Identification Number like a Social Security Number (SSN) or an FCC Registration.
- Must have a valid US Mailing Address

The Federal Communications Commission (FCC) is in charge of overseeing the Amateur Radio Service following the terms of the Communications Act of 1934. Furthermore, it is regulated by numerous international agreements. All operators of Amateur Radio stations must possess a license. In the US, licenses can be obtained in three main categories. The amount of frequencies that the user can access increases with the type of license. You have to pass an increasingly difficult test to get a license of a higher class.

License exams are given by volunteer groups of amateur radio operators, despite being under the FCC's jurisdiction. Under the supervision of organizations called Volunteer Examiner Coordinators, volunteers are in charge of conducting, grading, and reporting exam results to the FCC, the licensing body. US licenses are open to all holders, except for representatives of other governments, and are good for ten years before they must be renewed.

Choose your level of license

The Federal Communications Commission (FCC) provides three different categories of licenses to its customers (Federal Communications Commission, the branch of government that controls ham radio licenses). You must earn a passing score on an exam to move to higher levels, and you must begin at the Technician level.

In December 1999, following a thorough examination of the Amateur Radio licensing framework, the Federal Communications Commission (FCC) began enacting major revisions. In April 2000, the number of licensure classes decreased from six to three, which remains the current number. Furthermore, in February 2007, the Federal Communications Commission (FCC) discontinued requiring exams for proficiency in Morse code.

These brand-new laws were released by the Federal Communications Commission (FCC) to update the licensing procedure and bring amateur radio into the digital age. Licensed operators will be able to move more quickly from the novice to the expert levels of amateur radio, even though the new licensing structure may make it harder to get started.

Technician License

For nearly all new ham radio operators, the technician class license is the preferred entry-level license. You must pass an exam consisting of roughly 35 questions covering radio theory, regulations, and operating procedures to obtain this license. Upon obtaining the license, one can speak locally and frequently within North America on any amateur radio frequency over 30 megahertz. Additionally, it provides space for specific restricted rights on the high-frequency (HF; sometimes known as short wave) bands used for international communications.

General License

The General class license affords its holder certain operating privileges over the whole spectrum of Amateur Radio bands and modes of operation. This license makes it possible to communicate with people all around the world. To obtain a license for the General class, you will need to pass a test consisting of 35 questions. Those who hold a license for the General class must also have passed the written test for the Technician class.

Amateur Extra License

The Amateur Extra class license allows its holder to operate in all bands and modes within the Amateur Radio Service of the United States. The license is more difficult to obtain because you have to pass a fifty-question comprehensive test. Furthermore, candidates seeking an additional class license must provide proof that they passed the written examinations for each of the lower-level licensing classes.

GMRS License

There are several options to consider when selecting a communications service for a family or business. The General Mobile Radio Service, the Multi-Use Radio Service, the Family Radio Service, and the Citizens Band Radio Service (CBRS) are a few of these choices. These services are all offered (GMRS). Of these four services, the General Military Requirements System is regarded as the most capable. For example, even though many GMRS and FRS channels share the same frequencies, GMRS users can broadcast at a higher power level than FRS users, they can install repeaters to extend the range of their radios, and they can use certain limited data applications, like text messaging and GPS location.

But there is a price to be paid for possessing these abilities. While a license is not required to operate a CBRS, FRS, or MURS system, using a GMRS system requires obtaining one from the FCC, which costs $70 and is good for ten years. A single license allows all members of a family, regardless of age, to operate GMRS stations, however obtaining one requires you to be at least 18 years old (and you cannot represent the government of another nation).

If you do not already have an FCC Registration Number, often known as an FRN, you will need to obtain one before proceeding with the application process for a GMRS license.

- To begin, open the Commission's Registration System (CORES) in your web browser and then click the **REGISTER button**. Click the **CONTINUE button** once you have indicated if the FRN is for an individual or a corporation and whether the address for the contact is located within the United States. Fill out the information required for registration (the form can be found below), then click the **SUBMIT button**. After you have finished filling out the form, the CORES system will provide you with an FRN.

The next thing you need to do is apply for your GMRS license. To submit your license, go to the FCC License Management, enter your FRN and password, and then click the **SUBMIT button**.

You must click the **CONTINUE TO CERTIFY button** to be taken to the Certification page. To get a GMRS license, all you have to do is verify on this website that you meet the requirements. After entering your name and, if applicable, a title, click the **SUBMIT APPLICATION button**. A confirmation of your application will then show up in front of you. Click the **CONTINUE FOR PAYMENT OPTIONS** button to proceed with the payment process. You will be redirected to a page with multiple payment options, similar to the one below, after logging in with your FRN and password on the next screen. You can decide to make your payment with a credit card.

You will then be finished with the payment procedure after entering your payment details and selecting "**Continue with Plastic Card Payment**," which will bring you to the next screen. You will receive an email in due course verifying that your credit card was successfully charged and notifying you that a GMRS license has been obtained on your behalf. Re-log into the FCC website, go to My Licenses and select the option to Download Electronic Authorizations. This will enable you to print off or store a copy of the licensing authorization on your computer.

Activity

1. How can you get licensed to make use of the Baofeng radio?
2. What are the levels of the license listed by the FCC for ham radio operators?

CHAPTER 19
SOME RECOMMENDED ACCESSORIES

To converse efficiently, ham radio operators must have the proper equipment. Operators can enhance their signals, decrease noise, and extend their range with the right accessories. This chapter will teach you the top ten must-have accessories for any Baofeng or ham radio operator.

- **Antenna Tuner**: An antenna tuner is a unique, indispensable device that helps match the impedance that may be present between your radio and the antenna. It contributes to increased signal intensity and decreased noise, both of which boost performance significantly.

- **Antenna Analyzer**: this is a device that aids in the measurement of the performance of your antenna. It helps with the identification of any problems with your antenna and enables you to make necessary modifications for optimal performance.

- **Noise Canceling Headphones:** these headphones are very important accessories for any Baofeng or ham radio operator. They help with the reduction of outside noise, offer much better sound quality, and also give room for you to communicate without having to interfere with other people's communication.

- **SWR Meter:** The efficiency of your antenna system can be measured with the use of a standing wave ratio (SWR) meter. It takes care of identifying any issues with your antenna and handles the duty of implementing the changes required for the best results.

- **Dummy Load**: A dummy load is a device that simulates an antenna without having to send the signal. It's quite useful for testing as well as tuning your radio without having to broadcast a signal on the air.

- **Antenna Switch**: You can easily swap between different antennas with this switch. It helps save time and effort, which is especially helpful for operators with a variety of antennas.

- **A Computer**: This is another vital accessory for modern ham radio or Baofeng operators. It enables the running of digital modes, logs contacts, and also helps to gain access to online resources.

- **Mobile Antenna Mount**: For Baofeng radio operators, who frequently choose to operate from their cars, this device is rather crucial. It assists in giving you a convenient and safe option to install your antenna while you're out and about.

- **Ferrites**: Ferrites are tiny magnetic cores that are highly helpful in lowering electronic device interference. Noise reduction can be achieved by placing them on audio cables, power cords, and other lines.

- **Radio Amplifier**: a radio amplifier helps with the increment of the power output of your radio. It is very useful for weak signals or establishing communication over long distances.

Activity

1. Mention four (4) must-have accessories and their uses.

CHAPTER 20
A LOOK AT WINLINK

Winlink is a global radio messaging system that provides radio interconnection services using both government and amateur-band radio frequencies. Email with attachments, position reporting, weather updates, messaging about emergencies and relief efforts, and message relays are some of these services. Winlink goes by two names: the Winlink 2000 Network and Winlink Global Radio Email (a US service mark that is registered). The Amateur Radio Safety Foundation provides funding, and volunteers are in charge of setting up and maintaining the system.

Winlink networking (also known as ham radio) began with the provision of connectivity services for amateur radio. It is well known for being essential to the emergency and crisis communications networks of nations worldwide. In the past, the system used several central message servers spread throughout different regions of the globe to provide redundancy. But in 2017 and 2018, it was updated to use Amazon Web Services, which provides global content distribution, dynamic load balancers, and a geographically redundant cluster of virtual servers. As a component of the Winlink Hybrid Network, gateway stations have been using HF subbands for operations since 2013. If Internet connections are interrupted or rendered useless, this network offers message forwarding and delivery through a smart network that resembles a mesh. It was in the late 1990s and early 2000s that it began to evolve into what is today regarded as the global standard network infrastructure for amateur radio email. Additionally, the network was expanded to include separate parallel radio email networking systems for MARS, UK Cadet, the Austrian Red Cross, and the US Department of Homeland Security SHARES HF Program due to a need for improved communications during emergencies in the middle to later parts of the 2000s. This was done to meet the demand.

Participation in the Winlink system is permitted for amateur radio operators holding the required permits. The main beneficiaries of this system include emergency communications agencies, government and non-government public service groups, medical and humanitarian non-profits, and radio users without regular internet connection. The MARS component of the system is only available for usage by MARS operators who have received the necessary approval. As of July 2008, there were about 12,000 radio users and about 100,000 internet users who worked as correspondents. Around 100,000 communications are sent and received every month on average.

Winlink is occasionally used in place of or in addition to Sailmail, an amateur radio email system that uses marine HF frequencies instead of amateur ones, for sailors who want to cruise far from land. Winlink is used extensively. The service also uses a technology called Saildocs in addition to email. With Saildocs, sailors may send themselves links to resources on weather, sea safety, and other relevant subjects.

Winlink Express

Winlink Express (previously RMS Express) is the recommended Winlink radio email client since it is the only one that supports the Winlink Hybrid Network for email delivery with or without the internet. This is because it is the only client that supports the Winlink Hybrid Network and supports brand-new system features. With one or two additional IP addresses or alternate Winlink accounts, Winlink Express can be used to send and receive mail simultaneously. Winlink Express is made to be simple enough for lone users to utilize with just one call sign. It allows for an extensive selection of transceivers, TNCs, multimode controllers, sound card modes using the ARDOP, and VARA HF and FM virtual TNCs (ARDOP software TNC included), Pactor, SCS Robust Packet, VHF/UHF AX.25 packet radio, and direct telnet to CMS servers or RMS Relay (for amateur radio High-Speed Multimedia [HSMM], Broadband HamNet, D-Star DD mode, internet, and any other TCP/IP network).

The Winlink Development Team is in charge of creating Winlink Express, which utilizes Winlink's features while prioritizing usability in its design. It uses the open B2F extension radio transfer protocol, which supports multiple tactical addresses and attachments. Winlink Express has a peer-to-peer mode that allows users to connect to other Winlink Express or Airmail clients directly over radio frequency (RF). A propagation prediction feature that can help determine which of the participating Winlink RMS gateway HF stations is the best to connect to from anywhere in the world is another feature of Winlink Express. Other features include support for the Winlink catalog of downloadable weather, information, and help bulletins; Saildocs and Globalmarinenet for obtaining GRIB files; and both manual and automated GPS position reporting capabilities.

Winlink Express can be used as a client for emergency communications. It has EmComm-specific features such as HTML form creation and compact, formless content transport, as well as a growing library of automatically updated, included forms to use. Please inquire with your neighborhood EmComm group about their plans before adopting Winlink Express as your EmComm program. This could be your ARES, RACES, ACS, AuxComm, SHARES, or MARS organization that will be in charge of training and support.

System Requirements: Microsoft-compatible 32- or 64-bit Windows operating system (Windows Vista, Windows 7, 8, 10, or Windows 2003 Server, or later); it can also run Windows on Linux and Apple Mac computers by using a virtual machine engine or a dual boot configuration). Previous versions of Windows XP and above are not supported. The application uses little CPU power unless it is in a mode that uses a sound card modem. These modes need DSP, so a computer with a Pentium/Celeron CPU clocked at 700 MHz or higher and at least 2GB of RAM is required. It functions flawlessly on all contemporary PCs and Windows tablets. If multiple applications are running at the same time, a fast, late-model computer with 2-4GB RAM or more is recommended.

Winlink Express includes the following;

- **Built-in Customizable HTML Forms**: This uses form widgets, like selection boxes, checkboxes, and more, to allow data to be entered and seen on exact-likeness or simply "beautiful" forms within a browser. Furthermore, lightweight form contents can be sent over a network without the form itself, and when viewed through a browser on the receiving end, they can be automatically read inside the electronic form. Winlink Express installations include a large and regularly updated shared form template library of their own. Furthermore, it is possible to send the forms and templates as attachments over a radio link.

- **Built-in Message Log Report Generator**: This generates automatic reports based on the message records stored locally on your Winlink Express server. Produces ICS-309 reports for ICS compliance as well as text files with commas separated from each other. Provides you with the ability to select which logs should be included in the report (Inbox, Read items, Outbox, Sent items, Saved items, Drafts, Deleted items).

Message Entry

Winlink Express and Microsoft Outlook, as well as other email systems, share a lot of similarities. The user clicks on the corresponding icon to start writing a message. Subsequently, the text of the message is entered, and if necessary, one or more files are attached. Once the message is finished being created, it is saved in the "Outbox" until it is ready to be sent. The "Outbox," "Inbox," and other message storage folders are only a few of the databases that are saved on the user's computer. A message that has been moved to the Outbox is kept there indefinitely until the user starts a sending session. The message status is updated to "sent" once it has been sent, but the message is still stored in the database for further use.

Message Sending/ Receiving Sessions

The user will open one of the several session types when they are ready to send pending messages and check for incoming messages. Several other kinds of sessions can be had with Winlink Express, including telnet that does not use radio frequency (RF), Pactor, Packet, Winmor, ARDOP, Vara, and Iridium. When a session has been successfully opened, the user can begin the process of connecting to a server by clicking the **"Start" button.**

HF sessions like Pactor and Winmor come equipped with built-in busy-channel detectors that display a warning box whenever they pick up a signal on a frequency that is about to be used. Although this notification is helpful, before making a call, the user is supposed to listen to the audio on a frequency to identify whether or not a channel is busy.

Winlink Express uses the most appropriate transmission mode to start sending hailing calls as soon as the Start button is touched. The user's callsign and password will be authenticated through handshaking when the designated station responds to the call and is available. The station will not answer the call if it is unavailable. Any emails from the Outbox that are still awaiting delivery will be sent after this check. Any incoming communications for the callsign that are still pending are then forwarded to the session. The radio server connection will be cut off and transmission will cease once the session has completed sending and receiving messages.

Ham Operators and Winlink

There are many amazing things that ham radio operators are known for, such as using their radios to fly drones in first person, transmitting TV signals, and using a handheld transceiver to communicate with people all over the world. These are a few of the items that have become more and more popular in recent years. Another interesting thing amateur radio enthusiasts do is send emails via ham radio using a system called Winlink. This has proven to be a very useful addition to the amateur radio emergency communications toolbox. Winlink is a global communications system that operates automatically and without the Internet, using both government and amateur radio channels to enable email functionality. It is also known by its service mark, Winlink Global Radio Email, and as the Winlink 2000 Network, abbreviated as WL2K. This makes it possible for ham radio to take over when normal systems fail in times of crisis and emergency, while still being able to carry out a normal function like sending an email.

Winlink uses radio relays that are part of a smart network to send messages from one place to another. Emails sent through the Winlink system can upload attachments. Additionally, the system offers message relays, weather and information bulletins, emergency and relief communications, and position reporting through the Automated Position Reporting Service (APRS). To use the service, one must register as a Winlink operator. The operators follow a system that was created, maintained, and overseen by volunteer amateur radio operators. The Amateur Radio Safety Foundation provides system assistance; it is a nonprofit, noncommercial organization.

Winlink Account

Below are the steps you should follow to obtain a Winlink account; it is however worth noting that there is a need to possess a valid amateur radio license or a license from a participating government service or agency before attempting to obtain a Winlink account.

Note that ship station, marine, or general radiotelephone licenses do not qualify.

- Download, install, and also configure any client software. Winlink Express is the best for this process; you can also choose to study its help with installation and usage.

- If you opt for the use of Winkink Express, with an Internet connection, complete the form shown on the first startup after the installation has been completed. Ensure you add a password and password recovery address. Click on **Update**. This will make the Winlink Express process quite easy. With this, your account will be ready.

Make use of a different installation process apart from Winlink Express

To utilize and configure the program for your callsign, you must adhere to the guidance or directions given by the application. You must establish a connection with the system to set up your account (you can use Telnet to send a message). You shouldn't use a password when you first try to log in. For the radio, your email address is YOURCALL@winlink.org. In a few seconds, a letter with your password will be sent to your account. It can be obtained by utilizing an alternative connection. You must set your password within the client application because the CMS will require secure logins after you have recovered it.

Activity

1. Describe Winlink.

2. What are the various steps to owning a Winlink account?

Conclusion

You can start your ham radio pastime with a cheap device like the Baofeng. You can also use it as a trustworthy tool for networking and data backup if you're an expert. The Baofeng handheld device has a 50 Continuous Tone Coded Squelch System and a 104 Digital-Coded System display. These two systems have two bands. It also features a two-way radio system with the ability to shut out channels that are already in use and to choose between high and low decibels per milliwatt power.

A huge, crystal-clear LCD screen and a background light with three customizable colors are further highlights of this two-way radio. Danger signals, switchable radio frequency power, and LED flashlights with High and Low power settings are further features. These tools, which also come with a manual support program, make programming on a personal computer easier. Finally, but just as importantly, it works with CHIRP, a free programming tool designed mostly for radios.

In light of its reasonable cost and extensive feature set, Baofeng is difficult to dismiss. It is a versatile unit that can serve either as a backup unit for professionals or as a starter unit for novices who do not wish to invest a significant amount of money. It is ideal for a wide range of applications and works perfectly for any kind of user. Baofeng has a lot of value considering what it is and how much it costs.

Appendix A: Calling Frequencies

160 METERS

- 1.810 QRP CW Calling Frequency
- 1828.5 — DXpeditions CW Operations are frequently here
- 1.830-1.840 CW, RTTY, and other narrowband modes, intercontinental QSOs only
- 1.840-1.850 CW, SSB, SSTV, and other wideband modes, intercontinental QSOs only
- 1.825 – SSB QRP Calling Frequency
- 1910 – SSB QRP Calling Frequency

80/75 METERS

- 3.500-3.510 CW DX Window
- 3.505 DXpeditions CW is frequently here
- 3.560 QRP CW Calling Frequency
- 3.590 RTTY DX
- 3.790-3.800 SSB DX Window
- 3.710 QRP Novice/Tech CW Calling Freq
- 3.845 SSTV
- 3.885 AM Calling Frequency
- 3.799 DXpeditions SSB are frequently here
- 3.985 QRP SSB Calling frequency

40 METERS

- 7.000 – 7.010 CW DX Window
- 7037 Pactor Calling frequency
- 7.040 RTTY DX
- 7.040 QRP CW Calling Freq
- 7.050 XTAL Controlled Rigs
- 7.290 AM
- 7.065 DXpedition SSB USA split to 7.150 and above

- 7.005 DXpeditions CW are frequently here
- 7.110 QRP Novice/Tech CW Calling Frequency
- 7.171 SSTV
- 7.285 QRP SSB Calling frequency
- 7.290 AM Calling frequency

30 METERS

- 10.106 QRP CW Calling Frequency
- 10.110 — DXpeditions CW are frequently here

20 METERS

- 14.025 DXpedition CW Freq — Usually Split
- 14.060 QRP CW Calling Frequency
- 14.080 DXpedition RTTY Freq
- 14.080 to 14.100 Primary Range for RTTY

17 METERS

- 18.075 DXpeditions CW are frequently here — Usually Split
- 18.080 CW QRP Calling Freq
- 18.110 NCDXF Beacons (STAY OFF OF THIS FREQUENCY) Many Hams rely on these beacons for propagation determination.
- 18.130 SSB QRP Calling Freq
- 18.145 DXpeditions SSB are frequently here — Usually Split

15 METERS

- 21.025 Rare DX & DXpeditions Frequently Operate CW Here – Generally Listening Up-Split
- 21.060 QRP CW calling frequency
- 21.080 to 21.100 RTTY Primary Range
- 21.080 RTTY DXpeditions are frequently here
- 21.110 QRP Novice/Tech Calling Freq
- 21.150 NCDXF/IARU beacons (STAY OFF OF THIS FREQUENCY) Many Hams rely on these beacons for propagation determination.

- 21.295 Rare DX & DXpeditions Frequently Operate SSB Here — Generally Listening Up-Split
- 21.340, 21430 SSTV
- 21.385 QRP SSB calling frequency

10 METERS

- 28.025 CW Rare DX & DXpeditions Frequently Operate Here –Split
- 28.060 QRP CW Calling Frequency
- 28.080 RTTY Rare DX & DXpeditions Frequently Operate Here — Split
- 28.080 to 28.100 Primary Range for RTTY
- 28.1010 10/10 Intl CW Calling Frequency
- 28.110 QRP Novice/Tech Calling FREQ
- 28.190-28.225 Beacons

Appendix B: Bands for LEGAL Use

The amateur radio (HAM) bands are well-represented in the HF portion of the short-wave spectrum. These ham radio bands or frequency allocations are available for use by radio hams worldwide, although actual ham radio allocations do differ greatly between nations and regions. A high-level description of the ham radio band allocations is given first, followed by a discussion of the characteristics of the various radio amateur allocations.

There are nine unique frequency bands that ham radio operators share worldwide in the high-frequency (HF) section of the radio spectrum. There could be minor differences from one nation or area to another, depending on the particular amateur radio frequency in question. An explanation of each tier can be found below.

FZAmateur Radio Band (meters)	UK Allocation MHz	USA Allocation MHz
160	1.810 - 2.000	1.800 -2.000
80	3.500 -3.800	3.500 - 4.000
40	7.000 - 7.200	7.000 - 7.300
30	10.100 - 10.150	10.100 - 10.150
20	14.100 - 14.350	14.100 - 14.350

17	18.068 -18.168	18.068 - 18.168
15	21.000 -21.450	21.000 - 21.450
12	24.890 - 24.990	24.890 - 24.990
10	28.000 - 29.700	28.000 - 29.7000

Appendix C: GMRS/FRS Frequencies & Data

Operating in the United States, the General Mobile Radio Service (GMRS) is a mobile UHF two-way radio service. Users of this service must have a current license. Because they share the same frequencies, the General Mobile Radio Service (GMRS) and the Family Radio Service (FRS) can interact with each other. For the FRS, this frequency chart is also applicable.

The maximum power allowed for GMRS radios by the Federal Communications Commission (FCC) is 50 watts, a substantial increase above the 4 watts that are permitted for CB radios. The GMRS spectrum is divided into 22 channels: 22 shared channels with the FRS and 8 repeater channels reserved for GMRS use, which may only be accessed by licensed GMRS operators. GMRS operates on frequencies that range from 462.5625 to 467.7250.

Channel	Frequency(MHz)	Radio Service	Max Permitted Power -FRS	Max Permitted Power - GMRS
1	462.5625	GMRS/FRS	2 watts	5 watts
2	462.5875	GMRS/FRS	2 watts	5 watts
3	462.125	GMRS/FRS	2 watts	5 watts
4	462.6375	GMRS/FRS	2 watts	5 watts
5	462.6625	GMRS/FRS	2 watts	5 watts
6	462.6875	GMRS/FRS	2 watts	5 watts
7	462.7125	GMRS/FRS	2 watts	5 watts
8	467.5625	GMRS/FRS	.5 watts	.5 watts
9	467.5875	GMRS/FRS	.5 watts	.5 watts
10	467.6125	GMRS/FRS	.5 watts	.5 watts
11	467.6375	GMRS/FRS	.5 watts	.5 watts

12	467.6625	GMRS/FRS	.5 watts	.5 watts
13	467.6875	GMRS/FRS	.5 watts	.5 watts
14	467.7125	GMRS/FRS	.5 watts	.5 watts
15	462.5500	GMRS/FRS	2 watts	50 watts
16	462.5750	GMRS/FRS	2 watts	50 watts
17	462.6000	GMRS/FRS	2 watts	50 watts
18	462.6250	GMRS/FRS	2 watts	50 watts
19	462.6500	GMRS/FRS	2 watts	50 watts
20	467.6750	GMRS/FRS	2 watts	50 watts
21	462.7000	GMRS/FRS	2 watts	50 watts
22	462.7250	GMRS/FRS	2 watts	50 watts
15RP	467.5500	GMRS	Prohibited	50 watts
16RP	467.5750	GMRS	Prohibited	50 watts
17RP	467.6000	GMRS	Prohibited	50 watts
18RP	467.6250	GMRS	Prohibited	50 watts
19RP	467.6500	GMRS	Prohibited	50 watts
20RP	467.6750	GMRS	Prohibited	50 watts
21RP	467.7000	GMRS	Prohibited	50 watts
22RP	467.7250	GMRS	Prohibited	50 watts

Appendix D: MURS Frequencies

MURS, which stands for "Multi-Use Radio Service," is a group of five channels in the VHF band of the radio spectrum that does not require a license to operate. These channels can be used for both personal and commercial purposes. The United States has allowed the use of the following radio frequencies for MURS communications.

Channel	Frequency	Maximum authorized bandwidth	Channel name
1	151.82 MHz	11.25 kHz	MURS 1

2	151.88 MHz	11.25 kHz	MURS 2
3	151.94 MHz	11.25 kHz	MURS 3
4	154.57 MHz	20.00 kHz	Blue Dot
5	154.60 MHz	20.00 kHz	Green Dot

The use of channels 1–3 is required for "narrowband" FM, which has a deviation of 2.5 kHz and a bandwidth of 11.25 kHz. Both Channels 4 and 5 have the option of utilizing "wideband" FM, which has a deviation of 5 kHz and a bandwidth of 20 kHz, or "narrowband" FM.

Due to grandfathering provisions, business band radios that can use these frequencies are co-located with MURS channels 4-5. The "color dot system" used by commercial radio makers to differentiate between the different radio channels is how the "Blue Dot" and "Green Dot" radio channels received their names. Utilizing "privacy tones or codes" from CTCSS and DCS—which don't offer privacy—is allowed on MURS.

Appendix E: VHF Marine Frequencies

0	Ship Transmit MHz	Ship Receive MHz	Description of Communications
1A	156.050	156.050	Port Operations and Commercial, VTS.
5A	156.250	156.250	Port Operations or VTS.
6	156.300	156.300	Inter-ship Safety
7A	156.350	156.350	Commercial
8	156.400	156.400	Commercial (Inter-ship only)
9	156.450	156.450	Boater Calling. Commercial and non-commercial.
10	156.500	156.500	Commercial
11	156.550	156.550	Commercial. VTS in chosen areas
12	156.600	156.600	Port Operations. VTS in selected areas
13	156.650	156.650	Intership Navigation Safety (Bridge-to-bridge)

14	156.700	156.700	Port Operations. VTS in selected areas
15	—---		Environmental (Receive only).
16	156.800	156.800	International Distress, Safety, and Calling. Ships required to carry radio, USCG, and most coast stations maintain a listening watch on this channel
17	156.850	156.850	State Control
18A	156.900	156.900	Commercial
19A	156.950	156.950	Commercial
20	157.000	157.000	Port Operations
20A	157.000	157.000	Port Operations
21A	157.050	157.050	U.S. Coast Guard only
22A	157.100	157.100	USCG Liaison/Maritime Safety Information Broadcasts
23A	157.150	157.150	U.S. Coast Guard only
24	157.200	157.200	Public Correspondence (Marine Operator)
25	157.250	157.250	Public Correspondence (Marine Operator)
26	157.300	157.300	Public Correspondence (Marine Operator)
27	157.350	157.350	Public Correspondence (Marine Operator)
28	157.400	157.400	Public Correspondence (Marine Operator)
63A	156.175	156.175	Port Operations and Commercial, VTS
65A	156.275	156.275	Port Operations
66A	156.325	156.325	Port Operations
67	156.375	156.375	Commercial. Bridge-to-bridge communications in lower

			Mississippi River
68	156.425	156.425	Non-Commercial
69	156.475	156.475	Non-Commercial
70	156.525	156.525	Digital Selective Calling (voice communications not allowed)
71	156.575	156.575	Non-Commercial
72	156.625	156.625	Non-Commercial (Internship only)
73	156.675	156.675	Port Operations
74	156.725	156.725	Port Operations
77	156.875	156.875	Port Operations (Inter-ship only)
78A	156.925	156.925	Non-Commercial
79A	156.975	156.975	Commercial. Non-Commercial in Great Lakes only
80A	156.025	156.025	Commercial. Non-Commercial in Great Lakes only
81A	156.075	156.075	U.S. Government only - Environmental protection operations.
82A	156.125	156.125	U.S. Government only
83A	156.175	156.175	U.S. Government only
84	156.225	156.225	Public Correspondence (Marine Operator)
85	156.275	156.275	Public Correspondence (Marine Operator)
86	156.325	156.325	Public Correspondence (Marine Operator)
87A	156.375	156.375	Public Correspondence (Marine Operator)
88A	156.425	156.425	Commercial, Inter-ship only
AIS1	161.975	161.975	Automatic Identification System

| | | | (AIS) |
| AIS2 | 161.025 | 161.025 | Automatic Identification System (AIS) |

Appendix F: Frequency Coordinators

Association of Public-Safety Communications Officials, Inc. (APCO)

Automated Frequency Coordination Department

- 351 N. Williamson Blvd
- Daytona Beach, FL 32114-1112
- **phone:** 888-272-6911
- **fax:** 386-322-2502
- **email:** afc@apcointl.org

International Municipal Signal Association (IMSA)

- 122 Baltimore Street Suite 7
- Gettysburg, PA 17325
- **phone:** 717-398-0822
- **phone:** 855-803-1465
- **fax:** 717-778-4237
- **email:** wendy.jeffres@frequencycoordination.org

Forestry Conservation Communications Association (FCCA)

- 122 Baltimore Street Suite 7
- Gettysburg, PA 17325
- **phone:** 717-398-0815
- **phone:** 855-803-1465
- **fax:** 717-778-4237
- **email:** wendy.jeffres@frequencycoordination.org

American Association of State Highway and Transportation Officials (AASHTO)

- c/o RadioSoft

- 194 Professional Park Drive
- Clarkesville, GA 30523
- **phone:** 888-601-3676
- **fax:** 706-754-2745
- **email:** aashto@radiosoft.com

Forest Industries Telecommunications (FIT)

- 1565 Oak Street
- Eugene, Oregon 97401
- **phone:** 541-485-8441
- **phone:** 888-583-2929 Chicago
- **phone:** 888-342-2929 Dallas
- **phone:** 888-355-2929 Los Angeles
- **phone:** 888-395-2929 Washington, DC
- **fax:** 541-485-7556
- **email:** license@landmobile.com

Manufacturers Radio Frequency Advisory Committee, Inc. (MRFAC)

- c/o Radiosoft
- 194 Professional Park Drive
- Clarkesville, GA 30523
- **phone:** 800-262-9206
- **email:** coord@mrfac.com

Hydrological Federal Frequency/NOAA National Weather Service

- John Bradley
- Office of Climate, Water, and Weather Services, W/OS31
- 1325 East-West Highway, Room 13468
- Silver Spring, Maryland 20910
- **phone:** 301-427-9360
- **email:** hydro.radio.freq@noaa.gov or john.bradley@noaa.gov

American Association of State Highway and Transportation Officials (AASHTO)

- c/o RadioSoft
- 194 Professional Park Drive
- Clarkesville, GA 30523
- **phone:** 888-601-3676
- **fax:** 706-754-2745
- **email:** aashto@radiosoft.com

International Municipal Signal Association (IMSA)

- 122 Baltimore Street Suite 7
- Gettysburg, PA 17325
- **phone:** 717-398-0822
- **phone:** 855-803-1465
- **fax:** 717-778-4237
- **email:** wendy.jeffres@frequencycoordination.org

IMSA/FCCA/IAFC Frequency Coordination

- 122 Baltimore Street Suite 7
- Gettysburg, PA 17325
- **phone:** 717-398-0815
- **fax:** 717-778-4237
- **email:** wendy.jeffres@frequencycoordination.org

Forestry Conservation Communications Association (FCCA)

- 122 Baltimore Street Suite 7
- Gettysburg, PA 17325
- **phone:** 717-398-0815
- **phone:** 855-803-1465
- **fax:** 717-778-4237
- **email:** wendy.jeffres@frequencycooordination.org

Association of Public Safety Communications Officials, Inc (APCO)

- Automated Frequency Coordination Department
- 351 N. Williamson Blvd
- Daytona Beach, FL 32114-1112
- **phone:** 888-272-6911
- **fax:** 386-322-2502
- **email:** afc@apcointl.org

Appendix G: SWR Meters

One tool that aids in determining the standard wave ratio in a transmission line is the standing wave ratio meter, also known as the SWR meter, ISWR meter, or VSWR meter. The degree of mismatch between a transmission line and its load—which is frequently the antenna—is measured indirectly by the meter. Electronics professionals also use this instrument to assess the efficacy of prior impedance-matching attempts and adjust radio transmitters, together with their antennas and feedlines, to make sure they are impedance-matched and function properly.

Appendix H: CTCSS Squelch Tones (Hz)

The land mobile radio industry has been using the Continuous Tone Coded Squelch System, or CTCSS, since the late 1960s. This system goes by a more popular name. Although tone squelch is a general word, it is marketed under several different names, such as Motorola's Private Line (PL) and General Electric's Channel Guard (CG). This technology is produced by both businesses.

More than one agency (or fleet) can use the same radio frequency without being able to hear the other agency utilizing that frequency because sub-audible tones are transmitted with the spoken portion of the message. Sub-audible tones are used to achieve this. Agencies XYZ's receivers are configured to only release their audio squelch when the proper tone at the proper sub-audible frequency is transmitted.

CTCSS is more frequently employed by receivers as an additional squelch precaution (e.g., to prevent engine noise from breaking squelch) and by repeater systems to prevent noise or interference from making the repeater squawk incessantly. Agencies no longer share frequencies as frequently as they formerly did.

There were about 38 frequencies that were below the human hearing range when the land mobile industry initially started. Since then, it has grown to the more well-recognized 50. The 50 tones that are now in common use are listed here, albeit there isn't a single, accepted standard tone number assignment or code letter that goes with each tone.

Appendix I: DCS Codes

The abbreviation DCS stands for "Digital Coded Squelch." It is a collection of digital codes that function as filters to minimize interference caused by undesired transmissions originating from other persons or groups that are operating on the same channel or frequency. Interference Eliminator codes, Digital Private Line (DPL) codes, and Digital Quiet Talk (DQT) codes are a few of the names that are frequently used to refer to DCS codes.

Each DCS code may have a unique number assigned to it in the radio menu for selection and programming options on certain consumer and business radios that support them. These radios are suitable for both business and residential use. Some radios only list the codes themselves, leaving you with the choice to choose them by yourself. You should be aware that DCS codes ending in N are positive codes and codes ending in I are negative codes.

Appendix J: Phonetic Alphabet

Important information is spelled out using phonetic alphabets so that it can be comprehended in any context. Instead of saying "A B C," you say "Alpha Bravo Charlie." Certain letters, including D, T, and V, can blur together in noisy surroundings, making it challenging to tell them apart. For use in amateur radio communications, the International Telecommunication Union (ITU) devised a standard phonetic alphabet, which is mentioned below. Use of the NATO or International Aviation variants of this alphabet may result in slight spelling differences. To use ham radios, you need to commit this phonetic alphabet to memory.

Appendix K: Radio Horizon Antenna Heights

The radio horizon is said to be the locus point at which direct rays from an antenna are in tangent with the surface of the earth. Assuming the earth is a perfect sphere without the presence of an atmosphere, the radio horizon would be a circle. The radio horizon of the transmitting and receiving antenna can be combined to bring an increment in the effective communication range.

Appendix L: Weather Radio Frequencies

Public broadcast services, usually offered by government-owned radio stations, have a dedicated channel for weather forecasts and reports. Emergency weather bulletins are aired in between regular weather updates as necessary. A weather radio is intended to receive this kind of service. When a severe weather advisory is aired, the majority of weather radios contain an alerting mechanism that will either sound an alarm or switch to a pre-tuned weather channel if the user has the radio muted or tuned to another band. In the case of a natural disaster, a child abduction notice, or a terrorist attack, weather radio services may also broadcast non-weather related emergency information.

Typically, they use FM transmission on a certain band of very high frequencies (VHF). Though in certain places or during emergencies, you may be able to catch it on an AM or FM broadcast station, a terrestrial TV station, or a local public, educational, and government access (PEG) cable TV channel, you'll probably need a radio scanner or a weather radio receiver if you want to listen to a weather radio broadcast.

Frequency	WX Channel	Marine Channel	Radio Preset
162.400 MHz	WX2	36B	1
162.425 MHz	WX4	96B	2
162.450 MHz	WX5	37B	3
162.475 MHz	WX3	97B	4
162.500 MHz	WX6	38B	5
162.525 MHz	WX7	98B	6
162.550 MHz	WX1	39B	7
161.650MHz	WX#	21B	
161.750 MHz	WX#	23B	
161.775 MHz	WX#	83B	
162. 000 MHz	WX#	28B	ASM2
163.275 MHz	WX#	113B	

INDEX

1

10 Codes 61

A

A Computer 80
A/ B select key 5
ACCESSORIES 80
Accessory Jack 6
Amateur Extra License 78
Amateur Nets 44
Amateur radio repeater 37
An autopatch 40
Antenna Analyzer 80
Antenna Switch 80
Antenna Tuner 80
Antennas .... xii, 25, 26
ANTENNAS 25
Aperture 26
Appendix 87, 89, 90, 91, 92, 95, 98, 99
Array 26
Audio Quality 71

B

band...xi, 2, 6, 9, 11, 16, 19, 21, 29, 30, 31, 32, 33, 34, 38, 39, 40, 53, 59, 67, 70, 71, 73, 74, 75, 76, 82, 89, 91, 92, 99, 100
Band Plans 29
Baofeng xi, xii, xiii, 1, 3, 5, 6, 9, 10, 11, 13, 14, 16, 17, 22, 24, 27, 28, 35, 49, 50, 52, 53, 54, 56, 68, 69, 70, 71, 72, 79, 80, 86
BAOFENG .... i, 53, 66
Baofengs 1, 12, 24, 70
Battery 4, 23, 71
Battery Level Indicator 4
Broadcast relay station, re-broadcaster, or translator 37

C

cable .... xii, 16, 22, 36, 48, 49, 66, 67, 68, 69, 100
Cables 68
Carrier Operation 9
Cellular repeater 37
Channel 6, 9, 10, 20, 50, 51, 56, 90, 91, 98, 100
Channel (MR) mode 6, 9
Channel selection 6
CHIRP .... xi, xii, 17, 48, 49, 50, 51, 52, 86
Citizens' Band Radio 74
CLUB 63
Club or Topic 45
Coaxial Cable 67, 68
Communication 2, 59
Configuring backlight time-out 14
Configuring the power-on message 14

Cost of Ownership 70
CTCSS 10, 12, 30, 39, 53, 55, 92, 98
Customization 13

D

DCS 10, 12, 13, 53, 92, 99
Display 13, 14
DTMF 11
Dual Watch 9, 10, 11
Dummy Load 80
Durability and Reliability 71
DX 32, 33, 34, 45, 60, 74, 87, 88, 89

E

Emergency Frequency 17
Enhanced worker safety 71

F

Family Radio Service 3, 52, 73, 74, 78, 90
FCC RULES 29
First Transmission 39
Formal Operation 44
Free net 43
Frequency (VFO) mode 5, 6, 17
FRMS 52

G

General License 78
General Mobile Radio Service 2, 52, 73, 78, 90
GMRS 2, 3, 25, 50, 52, 70, 73, 74, 75, 78, 79, 90, 91
GRMS 52

H

Ham Radio 45, 46, 58, 64, 77
Ham Radio Nets 45, 46
HAND PROGRAMMING 53

I

Informal Operation 44
Installation 23

K

Keypad Lock 5

L

LEGAL .... ii, 77, 89

Lens 26
license.. 2, 3, 16, 29, 40, 52, 60, 64, 65, 70, 72, 73, 74, 75, 77, 78, 79, 85, 90, 91, 96
LICENSED 77
LINGO 57
Log periodic 27

M

Making a Call 6
Menu and function keys 5
menu shortcuts 8
Message Entry 84
Microstrip 27
Microwave relay 37
Mobile Antenna Mount 80
Multi-Use Radio Service 73, 78, 91

N

Net 42, 43, 44, 45
Numeric Keypad 5

O

Operation 43, 44, 56
Optical Communicator Repeater 36

P

Passive repeater 37
Phonetic Alphabet 60, 61, 99
Pound # key 5
Power and Volume 6
Power-on message 14
Procedure 12

Q

Q codes 58
QRL 58
QRP 58, 65, 87, 88, 89
QRS 59
QRT 59
QRU 58
QRV 59
QRZ 59
QSL 58
QSO 40, 59, 60, 64
QTH 59

R

Radio Club 64, 65
RADIO MENU 8
RADIO NETS 42
Radio Repeater 37
Reflector 26

Removal 23
Repeater 38, 39
repeaters .... xii, 11, 17, 30, 34, 36, 37, 38, 39, 41, 51, 53, 54, 56, 73, 74, 78
Repeaters .... xii, 30, 36, 37, 39, 40, 53, 54
REPEATERS 36

S

Scanning 9, 10
Scanning modes 9
Search Operation 9
Selective calling 12
Setting scanner mode 9
Shortcuts 8
Side key 1- CALL (Broadcast FM and Alarm) 4
Side Key 2 (Monitor and Flashlight) 4
Side Tip 49, 50
Star * key 5
Status LED 4
SWR Meter 80

T

Technician License 78
Telephone repeater 36
Time Operation 9
Traveling wave 27

U

UHF .... xi, 9, 19, 20, 21, 25, 37, 65, 83, 90

V

VFO/MR - mode key 5
VHF ...... xi, 9, 19, 20, 21, 30, 37, 52, 65, 70, 73, 83, 91, 92, 100
VHF & UHF .... xi, 19

W

Weather stations 50
Winlink .... xiii, 82, 83, 84, 85
WINLINK 82
Winlink Account 85
Winlink Express 82, 83, 84, 85
Wire 27

X

XYL 60

Y

YL 60

www.ingramcontent.com/pod-product-compliance
Lightning Source LLC
Chambersburg PA
CBHW062110220526
45471CB00010B/3676